Complex Numbers

A. NICOLAIDES, B.Sc., (Eng.), C.Eng.M.I.E.E.
Senior Lecturer and Course Tutor
at
SOUTH EAST LONDON COLLEGE
becomes known as
Lewisham College

Priv... s Ltd.

First Published in Great Britain 1990 by
Private Academic & Scientific Studies Ltd.

ISBN 1 872684 00 9

PRIVATE ACADEMIC & SCIENTIFIC STUDIES LTD.
11, Baring Road, London, SE12 0JP

P R E F A C E

This book, which is part of the G C E "A" level series in Pure Mathematics covers the specialised topic of complex numbers.

The "A" level series in Pure Mathematics is comprised of 10 books covering thoroughly the syllabuses of most boards. The books are designed to assist the student wishing to master the subject *Pure Mathematics*.

Each book deals extensively with a specialised topic with ample examples worked out in full.
The series is easy to follow with minimum help.

The *complex numbers* book, like all the books in the series, is divided into two parts. In Part 1, the theory of complex numbers is comprehensively dealt with, together with many worked examples and exercises.
A step by step approach is adopted in all the worked examples.
Part II of the book acts as a problem solver for all the exercises set at the end of each chapter in Part 1.

Topics essential to answering *complex numbers* exercises are also given special attention in this book. Such topics include trigonometry, binomial theorems, polynomial functions, and the circle and its general equation.

Thanks are due to the following examining bodies who have kindly allowed me to use questions from their examination papers:

Oxford	(O)
Cambridge	(C)
Associated Examining Board	(A E B)
	(W J E C)
Joint Matriculation Board	(J M B.)

CONTENTS

COMPLEX NUMBERS

PART I THEORY AND WORKED EXAMPLES OF COMPLEX NUMBERS.

PART I

THEORY AND WORKED EXAMPLES OF COMPLEX NUMBERS

1. DEFINITION OF A COMPLEX NUMBER

A complex number is a number which is not real. The square root of minus one, that is, $\sqrt{-1}$ is a complex number because there is no real number which can be multiplied by itself in order to give the answer of -1.

The square roots of four, $\sqrt{4}$, however, are equal to ± 2 which are real numbers, because $2 \times 2 = 4$ or $(-2) \times (-2) = 4$.

Let us now examine the quadratic equation which has a negative discriminant:

$\mathcal{D} = b^2 - 4ac$ = discriminant, the quantity under the square root.

Solve the quadratic equation $x^2 + x + 1 = 0$

Applying the formula $x = \dfrac{-b \pm \sqrt{b^2 - 4ac}}{2a}$

we have $x = \dfrac{-1 \pm \sqrt{1^2 - 4(1)(1)}}{2 \times 1} = \dfrac{-1 \pm \sqrt{-3}}{2}$

One root is $\alpha = -\dfrac{1}{2} + \dfrac{\sqrt{-3}}{2}$ and the other root is $\beta = -\dfrac{1}{2} - \dfrac{\sqrt{-3}}{2}$, the roots are complex but the sum of the roots, $\alpha + \beta = -1$ and the product of the roots, $\alpha\beta = 1$.

The discriminant is negative, and therefore in the set of real numbers the above equation has no solution.

Leonhard EULER* and Karl Friedrich GAUSS** have extended the set of real numbers so that quadratic equations with negative discriminant can be solved.

The set of real numbers was extended to the set of complex numbers so that the set of real numbers is a proper subset of the set of complex numbers.

Mathematicians substituted $\sqrt{-1}$ by the letter i and Engineers substituted $\sqrt{-1}$ by the letter j. The letter i is the first letter of the French term 'imaginaire' which translated into English means 'imaginary'. Engineers use the j notation in order to avoid confusion with the letter 'i' for 'intensite' (which means 'current' in French).

$\sqrt{-3}$ can be written as $\sqrt{-3} = \sqrt{3}\sqrt{(-1)} = \sqrt{3}i$ in the roots of the above quadratic equation, therefore

$$\alpha = -\frac{1}{2} + \frac{\sqrt{3}}{2}i \text{ and } \beta = -\frac{1}{2} - \frac{\sqrt{3}}{2}i$$

These roots are complex, which are made up of two components, the real term $-\frac{1}{2}$, and the imaginary terms $+\frac{\sqrt{3}}{2}i$ and $-\frac{\sqrt{3}}{2}i$

It should be observed that the terms $\frac{\sqrt{3}}{2}i$ and $-\frac{\sqrt{3}}{2}i$ do not mean that i is multiplied by either $\frac{\sqrt{3}}{2}$ or $-\frac{\sqrt{3}}{2}$.

The term $\frac{\sqrt{3}}{2}i$ means that $\frac{\sqrt{3}}{2}$ is represented along the positive y-axis (imaginary axis) and $-\frac{\sqrt{3}}{2}i$ means that $\frac{\sqrt{3}}{2}$ is represented along the negative y-axis (imaginary axis).

* Leonard EULER was a Swiss mathematician born on 15th April 1707 in Basle and died on 18th September 1783 in St.Petersburg.

** Karl Friedrich GAUSS was a German mathematician born on 30th April 1777 in Brunswick and died on 23rd February 1855 in Gottingen. He was reputed to be one of the greatest mathematicians in Europe.

The following worked examples will illustrate the significance of a complex number or the negative discriminant.

WORKED EXAMPLE 1

Determine whether the straight line graph $y = x + 3$ is a tangent or intersects the parabola $y^2 = x$.

SOLUTION 1

Solving the simultaneous equations in x

$y^2 = x$ and $y = x + 3$, we have $(x + 3)^2 = x$ or $x^2 + 6x + 9 - x = 0$

or $x^2 + 5x + 9 = 0$.

The discriminant of this equation is negative,

$D = b^2 - 4ac = (5)^2 - 4(1)(9) = 25 - 36 = -9$,

this implies that the straight line neither touches the curve nor intersects it.

Solving the quadratic equation, we have that the roots are $\alpha = -\frac{5}{2} + \frac{3}{2}i$ and $\beta = -\frac{5}{2} - \frac{3i}{2}$ which are complex roots.

WORKED EXAMPLE 2

Write down the following numbers in complex number notation:

(i) $\sqrt{-5}$ (ii) $-3 - \sqrt{-3}$ (iii) $4 - \sqrt{-7}$ (iv) -2

SOLUTION 2

(i) $\sqrt{-5} = \sqrt{5}\,\sqrt{-1} = \sqrt{5}i$

(ii) $-3 - \sqrt{-3} = -3 - \sqrt{3}i$

(iii) $4 - \sqrt{-7} = 4 - \sqrt{7}i$

(iv) $-2 = -2 + 0i$

EXERCISES (1)

1. Write the following in complex number notation:

   (i) $\sqrt{-2}$ (ii) $\sqrt{-4}$ (iii) $\sqrt{-8}$ (iv) $\sqrt{-16}$ (v) $\sqrt{-27}$

   (vi) $1 + \sqrt{-3}$ (vii) $-1 - \sqrt{-5}$ (viii) $-5 + \sqrt{-7}$

2. Determine whether the following quadratic equations have real or complex roots:

   (i) $3x^2 - x + 1 = 0$

   (ii) $-x^2 + x - 5 = 0$

   (iii) $-5x^2 + 7x + 5 = 0$

   (iv) $x^2 - 4x + 8 = 0$

   (v) $x^2 + 2x + 2 = 0$

3. Find the complex roots of the quadratic equations in (2) above, and observe the relationship between the roots.

4. Determine whether the graphs intersect:

   (i) $3x - y + 1 = 0$ and $x^2 + y^2 = 1$

   (ii) $x^2 = 4y$ and $-x^2 = 4y$

   (iii) $x^2 = 4y$ and $x - y = 3$

   (iv) $x^2 + (y - 1)^2 = 1$ and $y = -3x + 4$

2. ## PLOTTING COMPLEX NUMBERS IN AN ARGAND DIAGRAM

THE QUADRATIC OR CARTESIAN FORM

Jean Robert ARGAND was a Swiss mathematician born in Geneva in 1768 and died in Paris in 1822. He employed complex numbers to show that all algebraic equations have roots.

CARTESIAN FORM OF A COMPLEX NUMBER

Cartesian form of a complex number is $Z = x + iy$ where Z is any complex number and x and y are real numbers, $x,\ y \in \mathbb{R}$.

The real part of Z is denoted $ReZ = x$ and the imaginary part of Z is denoted by $ImZ = y$.

$$ReZ = x$$
$$ImZ = y$$

René DESCARTES was a French philosopher and mathematician born on March 31st 1556 at La Haye Touraine and died on February 11th, 1650 in Stockholm. He is famed for his coordinate geometry or cartesian geometry.
Complex numbers can be represented in a diagram called 'The Argand Diagram' which is an extremely useful diagram in understanding complex numbers.

There are two cartesian axes, the x-axis which is the real axis and the y-axis which is the imaginary axis. These two perpendicular axes intersect at a point O, which is called the origin.
Fig.1 illustrates the Argand diagram.

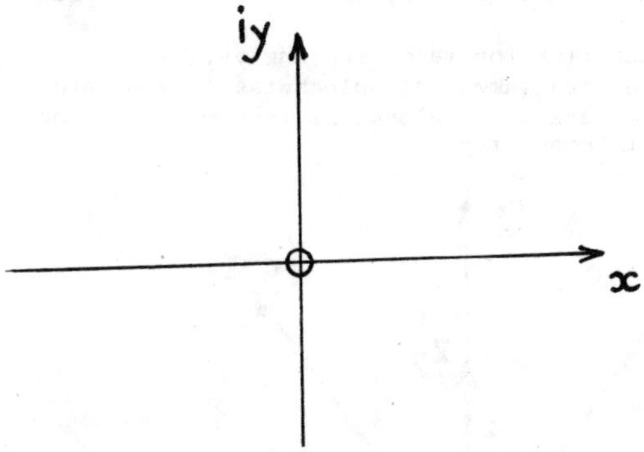

Fig.1 Cartesian axes Argand diagram.

WORKED EXAMPLE 3

(a) Plot the following numbers in an Argand diagram:

(i) $Z_1 = 3$
(ii) $Z_2 = -2$
(iii) $Z_3 = 3 + i4$
(iv) $Z_4 = 3 - i4$
(v) $Z_5 = -3 + i4$
(vi) $Z_6 = -3 - i4$
(vii) $Z_7 = 3i$
(viii) $Z_8 = -2i$
(ix) $Z_9 = 5 + i$
(x) $Z_{10} = -4 + i2$

(b) Express the above numbers in coordinate set form or in ordered pairs.

SOLUTION 3

(a) It is noted that some of the numbers are real and some are complex. Usually if Z is a complex number then $y \neq 0$ and x, $y \in \mathbb{R}$. If $y = 0$ then the number is real.

(i) Referring to Fig. 2, Z_1 is wholly real and is three units along the positive x-axis.

(ii) Z_2 is wholly real and is two units along the negative x-axis.

(iii) $Z_3 = 3 + i4$, is marked as follows:
three units along the real positive x-axis and
four units along the imaginary positive y-axis,
completing the parallelogram, the diagonal gives the vector Z_3.

(iv) Similarly $Z_4 = 3 - i4$, three units along the real positive x-axis, and four units along the negative imaginary y-axis, the diagonal of the parallelogram gives the vector Z_4.

(v) $Z_5 = -3 + i4$, three units along the negative real x-axis and four units along the positive y-axis thus forming a parallelogram whose diagonal is the vector Z_5.

(vi) Similarly $Z_6 = -3 - i4$ is plotted.

(vii) Z_7 is wholly imaginary, which is three units along the positive y-axis.

(viii) Z_8 is also wholly imaginary, which is two units along the negative imaginary axis.

(ix) $Z_9 = 5 + i1$, five units along the positive x-axis and one unit along the positive y-axis.
Z_9 is the diagonal of the parallelogram.

(x) $Z_{10} = -4 + i2$, four units along the negative real axis and two units along the posive y-axis, completing the parallelogram gives the diagonal which is the vector Z_{10}

All the above numbers are vectors, that is, they have magnitude and direction.
OX is the reference line for measuring angles.
The positive angles are taken anti-clockwise from OX and the negative angles are taken clockwise from OX as in the unit radius circle in trigonometry.

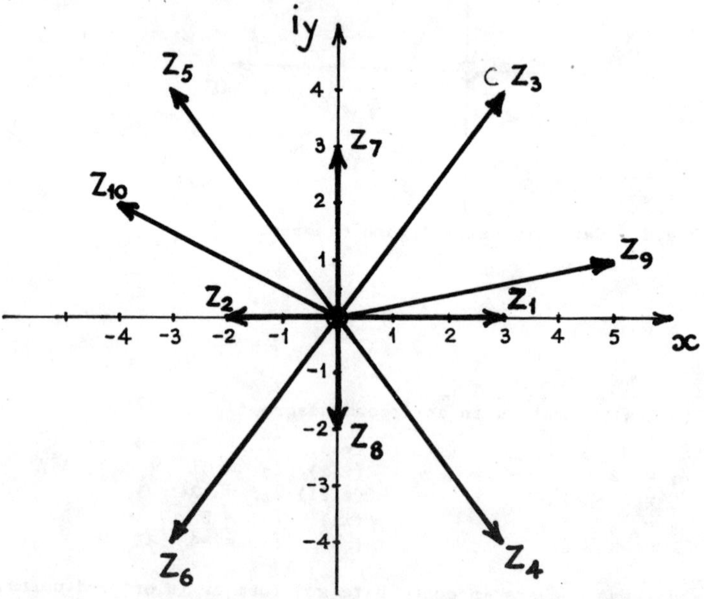

Fig.2 Complex numbers plotted on an Argand diagram.

(b) (i) (3,0) (iii) (3,4) (v) (-3,4) (vii) (0,3) (ix) (5,1)
(ii) (-2,0) (iv) (3,-4) (vi) (-3,-4) (viii) (0,-2) (x) (-4,2)

The coordinates of the point C are (3,4), that is:
three units along the x-axis and four units along the y-axis.
OC represents the complex number Z_3.

THE POWERS OF i

WORKED EXAMPLE 4

Represent the following complex numbers in an Argand diagram:

(i) i^5 (iv) i^{31}
(ii) i^9 (v) i^{38}
(iii) i^{25} (vi) i^{1985}

SOLUTION 4

$Z = 1$ is represented along the positive x-axis.

If the vector $Z = 1$ is rotated in an anticlockwise direction of 90°, the vector is i, this is obtained by merely multiplying the unity vector 1 by i. Similarly if the vector i is multiplied by i again, it results in vector i^2 or -1; if -1 is multiplied by i it becomes $-i$ or i^3 and the vector is now in the negative imaginary axis, and if $i^3 \times i = i^4 = 1$, we are back in the original direction, the positive x-axis. Therefore, by multiplying a vector by i, the vector is rotated through 90° in an anticlockwise direction with centre the origin O.

Complex numbers are vectors, i.e. they have magnitude and direction.

(i) $i^5 = i$, this is obtained by dividing 5 by 4, giving one complete revolution and leaving 1 as the remainder, which is further rotated by 90°, and i^5 is along the positive imaginary axis.

(ii) $i^9 = i$, this is obtained by dividing 9 by 4, giving two complete revolutions and $\frac{1}{4}$ of revolution.

(iii) $i^{25} = i$, this is obtained by dividing 25 by 4, giving six complete revolutions, leaving 1 as the remainder, i.e. a further 90° in anticlockwise direction.

(iv) $i^{31} = i^3 = -i$, this is obtained by dividing 31 by 4, giving 7 complete revolutions, leaving 3 as the remainder, i.e. $3 \times 90^\circ = 270^\circ$ in an anticlockwise direction.

(v) $i^{38} = i^2 = -1$

(vi) $i^{1985} = i$, this is obtained by dividing 1985 by 4, giving 496 complete revolutions and one quarter of a revolution in an anticlockwise direction.

Fig. 3 illustrates the above complex numbers.

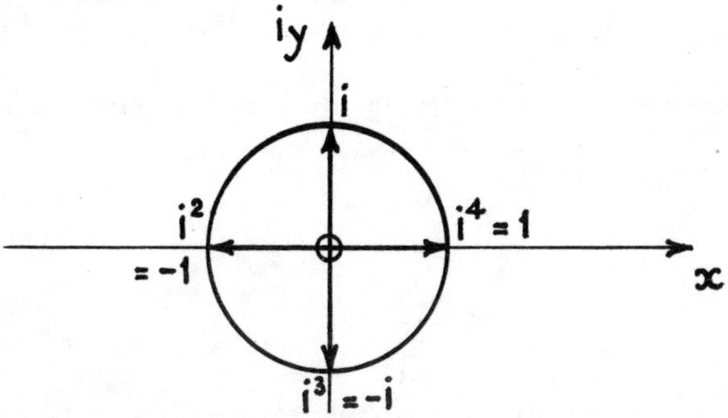

Fig.3 The powers of i, $i^2 = -1$, $i^4 = 1$, $i^3 = -i$ $i^{1985} = i$.

EXERCISES (2)

1. Express the following points of coordinates in the complex number form:

(i) A(1,3)	(vi) F(2,-4)
(ii) B(2,5)	(vii) G(0,0)
(iii) C(0,6)	(viii) H(a,b)
(iv) D(3,0)	(ix) I(x,y)
(v) E(-1,3)	(x) J(-3,-4)

2. Express the following complex numbers in the form of points of coordinates:

(i) $Z_1 = 3 + i4$	(viii) $Z_8 = -i - 2$
(ii) $Z_2 = 3 - i4$	(ix) $Z_9 = ai + b$
(iii) $Z_3 = -3 + i4$	(x) $Z_{10} = 7i$
(iv) $Z_4 = -3 - i4$	(xi) $Z_{11} = 3 - 2i$
(v) $Z_5 = 3i$	(xii) $Z_{12} = x - iy$
(vi) $Z_6 = -i$	(xiii) $Z_{13} = \cos\theta + i\sin\theta$
(vii) $Z_7 = -3$	

3. Plot the complex numbers in (2) in an Argand diagram.

4. The square root of (-1) is denoted by the letter i, i.e. $i = \sqrt{-1}$. Explain the meaning of i with the aid of an Argand diagram and hence simplify the following in terms of i

(i) i^2	
(ii) i^5	(iv) i^{33}
(iii) i^7	(v) i^{1986}

5. A complex number is a vector. Explain clearly the meaning of a vector by illustrating in an Argand diagram.

3 THE SUM AND DIFFERENCE OF TWO COMPLEX NUMBERS

There are two methods in determining the sum and difference of two complex numbers, the algebraic method and the graphical method, using the Argand diagram.

Determining the sum and difference of the complex numbers algebraically:

If $Z_1 = x_1 + iy_1$ and $Z_2 = x_2 + iy_2$
then the sum $Z_1 + Z_2 = (x_1 + iy_1) + (x_2 + iy_2) = (x_1 + x_2) + i(y_1 + y_2)$
and the difference $Z_1 - Z_2 = (x_1 + iy_1) - (x_2 + iy_2) = (x_1 - x_2) + i(y_1 - y_2)$.

The real terms are added or subtracted and the imaginary terms are added or subtracted separately.

WORKED EXAMPLE 5

Represent the complex numbers in an Argand diagram, and find their sum and difference:

$$Z_1 = 4 + i \quad \text{and}$$
$$Z_2 = 1 + i3.$$

SOLUTION 5

The complex numbers Z_1 and Z_2 are plotted in an Argand diagram in Fig.4.

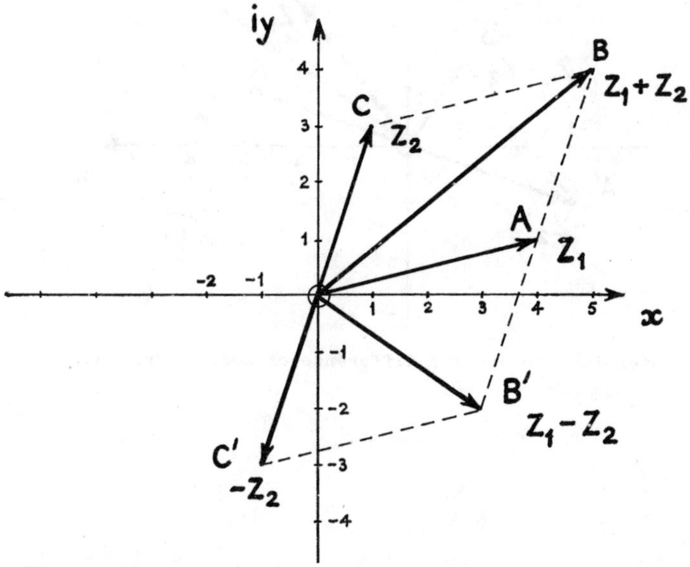

Fig.4 The sum and difference of complex numbers in an Argand diagram.

The resultant of the two vectors Z_1 and Z_2 is obtained by drawing the parallelogram $OABCO$. OB is the resultant

$$Z_1 + Z_2 = (4 + i) + (1 + i3) = 5 + i4$$

$$Z_1 + Z_2 = 5 + i4$$

To determine the difference of the two complex numbers, OC is projected in the opposite direction $OC' = -Z_2$.

The resultant of Z_1 and $-Z_2$ is obtained again by completing the parallelogram $OAB'C'O$. $OB' = Z_1 - Z_2 = (4 + i) - (1 + i3) = 3 - i2$

$$Z_1 - Z_2 = 3 - i2.$$

WORKED EXAMPLE 6

Find the sum and difference of the two complex numbers $Z_1 = 2 + i5$ and $Z_2 = 3 + i2$

(i) algebraically, and
(ii) graphically.

SOLUTION 6

(i) $Z_1 + Z_2 = (2 + i5) + (3 + i2) = (2 + 3) + i(2 + 5) = 5 + i7$
$Z_1 - Z_2 = (2 + i5) - (3 + i2) = (2 - 3) + i(5 - 2) = -1 + i3$

The real terms are added or subtracted and the imaginary terms are added or subtracted separately.

(ii) Z_1 and Z_2 are plotted in Fig.5 in the Argand diagram. From the diagram, $Z_1 + Z_2 = 5 + i7$ and $Z_1 - Z_2 = -1 + i3$ which agree with the above results.

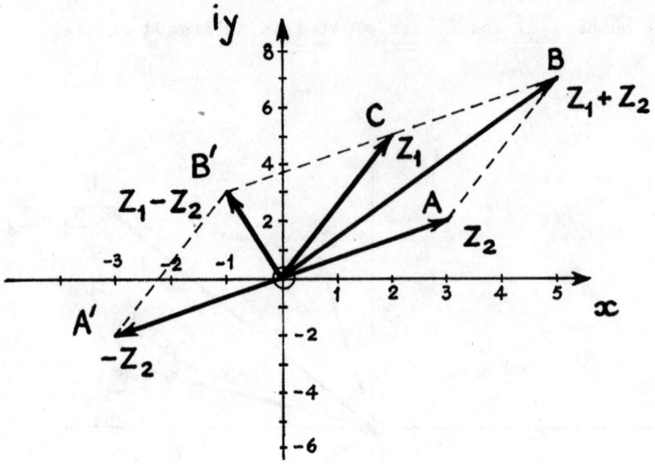

Fig.5 To determine the sum and difference of complex numbers.

EXERCISES (3)

1. If $Z_1 = 2 + 3i$, $Z_2 = 3 + 4i$, $Z_3 = -4 - i5$,
   determine the following complex numbers algebraically, expressing them in the form $a + ib$:

(i) $Z_1 + Z_2$	(v) $Z_1 - Z_3$	(ix) $Z_3 - 3Z_1$
(ii) $Z_1 + Z_3$	(vi) $Z_3 - Z_2$	(x) $Z_3 - 2Z_1$
(iii) $Z_2 + Z_3$	(vii) $2Z_1 + 3Z_3$	(xi) $Z_3 + 5Z_2$
(iv) $Z_1 - Z_2$	(viii) $Z_1 + 2Z_2$	

2. (i) If $ReZ = x$ and $ImZ = y$, write down the value of Z.

   (ii) If $ReZ = -3$ and $ImZ = 5$, write down the value of Z.

   (iii) If $ReZ = a$ and $ImZ = -b$, write down the value of Z.

3. Find the sum and difference of the vectors $E_1 = 20 + i30$ and $E_2 = 10 + i15$

4. On the same diagram, draw the vectors which represent the complex numbers $-3 + i2$ and $2 + i3$ respectively.
   Prove from your figure that the vectors are perpendicular.

5. (a) Determine the resultant of the two vectors $Z_1 = -3 + i2$ and $Z_2 = 2 + i3$

   (b) Determine the difference of the two vectors $Z_2 = -3 + i2$ and $Z_1 = 2 + i3$

4 DETERMINES THE PRODUCT OF TWO COMPLEX NUMBERS IN THE QUADRATIC FORM $Z = x + iy$

If $Z_1 = x_1 + iy_1$ and $Z_2 = x_2 + iy_2$

The product $Z_1Z_2 = (x_1 + iy_1)(x_2 + iy_2) = x_1x_2 + iy_1x_2 + ix_1y_2 - y_1y_2$

$$Z_1Z_2 = (x_1x_2 - y_1y_2) + i(y_1x_2 + y_2x_1)$$

$$Re(Z_1Z_2) = x_1x_2 - y_1y_2$$

$$Im(Z_1Z_2) = x_1y_2 + y_1x_2$$

WORKED EXAMPLE 7

Find the product of the following complex numbers

$$Z_1 = 3 + i4 \text{ and } Z_2 = 1 - i5$$

Plot Z_1, Z_2 and Z_1Z_2 in an Argand diagram.

SOLUTION 7

$Z_1Z_2 = (3 + i4)(1 - i5) = 3 - i15 + i4 - i^2 20 = 3 - i11 - (-20)$
$= 3 - i11 + 20 \qquad = 23 - i11$

where $i^2 = -1$

$Re(Z_1Z_2) = 23$

$Im(Z_1Z_2) = -11$

Fig. 6 shows Z_1, Z_2 and Z_1Z_2 in an Argand diagram.

Multiplication is defined as

$$(x_1, y_1) \odot (x_2, y_2) = (x_1x_2 - y_1y_2,\ x_1y_2 + x_2y_1)$$

Let $i = (0,1)$

then $i^2 = (0,1) \odot (0,1) = (0.0 - 1.1,\ 0.1 + 0.1) = (-1,0)$

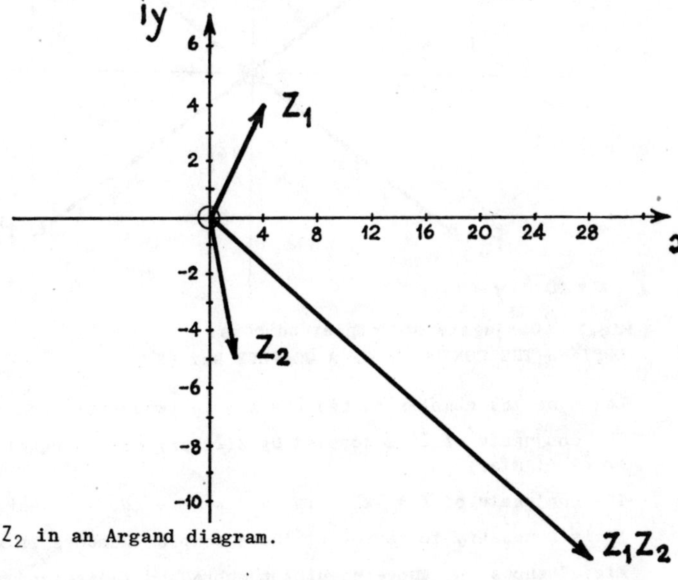

Fig.6 Z_1, Z_2, Z_1Z_2 in an Argand diagram.

EXERCISES (4)

1. Express the following basic operations in the form $a + ib$
   if $Z_1 = 3 - 4i$
   $Z_2 = 1 + i$
   $Z_3 = 2 + 3i$

   (i) Z_1Z_2
   (ii) Z_1Z_3
   (iii) Z_2Z_3
   (iv) $Z_1Z_2Z_3$

2. Express in the form $a + ib$

   (i) $(3i)(5i)$
   (ii) $(2 + 3i)(3 + 4i)$
   (iii) $(3 - 5i)(3 + 4i)$
   (iv) $(4 - 5i)(1 + i)$
   (v) $(1 + 2i)3$
   (vi) $5i(1 - i)$
   (vii) $(1 + i)(1 - i)$
   (viii) $(1 + 2i)(1 - 2i)$
   (ix) $(1 - 3i)(1 + 3i)$
   (x) $(4 + 3i)^2$
   (xi) $(a + bi)^2$
   (xii) $(\cos\theta + i\sin\theta)(\cos\phi + i\sin\phi)$
   (xiii) $(1 + 3i)^3$
   (xiv) $(1 - i)^3$
   (xv) $(1 - i^2)^5$

3. If $Re(Z_1Z_2) = x_1x_2 - y_1y_2$

   $Im(Z_1Z_2) = x_1y_2 + y_1x_2$

   Find Z_1Z_2.

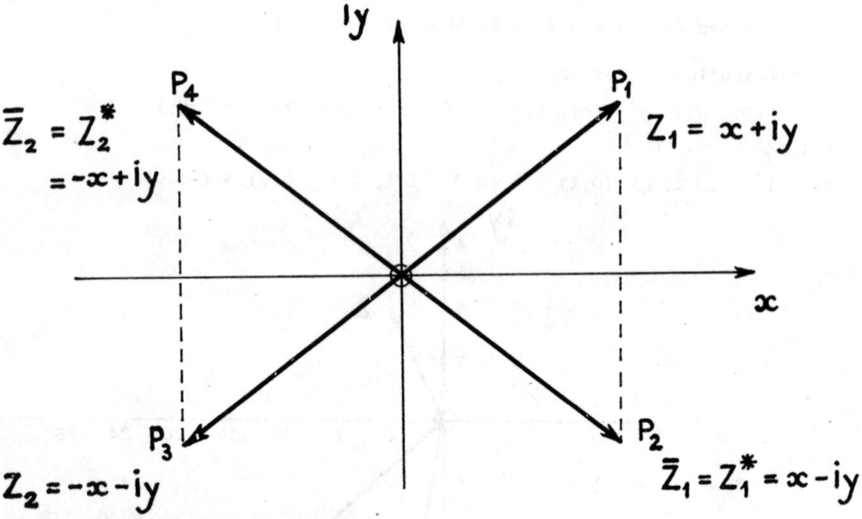

Fig.7 Conjugate of complex numbers.

5 DEFINES THE CONJUGATE OF A COMPLEX NUMBER.

Let Z be the complex number $Z = x + iy$ where $(x, y \in \mathbb{R})$.

The conjugate of Z is denoted by $\overline{Z}$ (Z bar) and is equal to $\overline{Z} = x - iy$ or Z^* (Z star).

The conjugate of $Z = -x - iy$ is $\overline{Z} = -x + iy$.

It is necessary to represent these complex numbers in an Argand diagram.

Fig. 7 shows the above complex numbers and their conjugates.

The reflection in the x-axis of P_1 is P_2, which is the conjugate of Z.

The conjugate of $Z_2 = -x - iy$ is again the reflection in the x-axis which is represented as $\overline{Z_2} = -x + iy$.

Note that the real quantity is unaltered, the imaginary term changes the sign.

When the complex number is expressed as a quotient which contains i in the denominator, it is necessary to multiply a quotient complex expression with the conjugate of the denominator in order to obtain a real quantity in the denominator.

The product of two conjugate numbers is always real and positive:

$$(x + iy)(x - iy) = x^2 + y^2$$

$$(-x - iy)(-x + iy) = (-x)^2 - i^2y^2 = x^2 + y^2$$

To prove that $\overline{Z_1 + Z_2} = \overline{Z_1} + \overline{Z_2}$ where $Z_1 = a_1 + ib_1$ and $Z_2 = a_2 + ib_2$

Proof:

$$\overline{Z_1 + Z_2} = \overline{(a_1 + a_2) + (b_1 + b_2)i}$$

$$= (a_1 + a_2) - (b_1 + b_2)i$$

$$= (a_1 - b_1 i) + (a_2 - b_2 i)$$

$$\overline{Z_1 + Z_2} = \overline{Z_1} + \overline{Z_2}$$

where $\overline{Z_1} = a_1 - ib_1$ and $\overline{Z_2} = a_2 - ib_2$.

To prove that $\boxed{\overline{Z_1 . Z_2} = \overline{Z_1} . \overline{Z_2}}$

Proof:

$$\overline{Z_1 Z_2} = \overline{(a_1 a_2 - b_1 b_2) + (a_1 b_2 + a_2 b_1)i}$$

$$= (a_1 a_2 - b_1 b_2) - (a_1 b_2 + a_2 b_1)i$$

$$\overline{Z_1} . \overline{Z_2} = (a_1 - b_1 i).(a_2 - b_2 i)$$

$$= \{a_1 a_2 - b_1 b_2\} + \{a_1(-b_2) + a_2(-b_1)\}i$$

$$= (a_1 a_2 - b_1 b_2) - (a_1 b_2 + a_2 b_1)i$$

then $\overline{Z_1 . Z_2} = \overline{Z_1} . \overline{Z_2}$

To prove that $\boxed{\overline{\left(\frac{Z_1}{Z_2}\right)} = \frac{\overline{Z_1}}{\overline{Z_2}}}$, $Z_2 \neq 0$

If $Z_1 = 1$ $\quad Z_2 = Z$ then $\overline{\left(\frac{1}{Z}\right)} = \frac{1}{\overline{Z}}$

If $Z = \frac{Z_1}{Z_2}$ when $Z_2 \neq 0$

then $ZZ_2 = Z_1$ $\quad \overline{ZZ_2} = \overline{Z_1}$ $\quad \overline{Z}.\overline{Z_2} = \overline{Z_1}$ $\quad \overline{Z} = \overline{\left(\frac{Z_1}{Z_2}\right)}$

then $\overline{\left(\frac{Z_1}{Z_2}\right)} = \frac{\overline{Z_1}}{\overline{Z_2}}$

It is proved by induction that $\overline{Z_1 + Z_2 + \ldots\ldots + Z_n} = \overline{Z_1} + \overline{Z_2} + \ldots + \overline{Z_n}$

$\overline{Z_1 Z_2 \ldots\ldots\ldots\ldots Z_n} = \overline{Z_1}\,\overline{Z_2} \ldots\ldots\ldots \overline{Z_n}$

If $Z_1 = Z_2 = \ldots = Z_n = Z$ then $\overline{(Z^n)} = (\overline{Z})^n$

If $\overline{a + bi} = a - bi$ then $(\overline{\overline{Z}}) = Z$

$Z + \overline{Z} = 2a$ a real number

$Z - \overline{Z} = 2bi$ an imaginary number

$Z\overline{Z} = a^2 + b^2$ a real number

$\overline{Z_1 + Z_2} = \overline{Z_1} + \overline{Z_2}$

6 DETERMINES THE QUOTIENT OF TWO COMPLEX NUMBERS.

Let Z be the quotient of two complex numbers $Z = \dfrac{x_1 + iy_1}{x_2 + iy_2}$ ....(1)

It is required to express the complex numbers in the form $a + ib$.

Multiplying numerator and denominator of equation (1) by the conjugate of $x_2 + iy_2$, namely $x_2 - iy_2$, we have

$$Z = \frac{(x_1 + iy_1)}{(x_2 + iy_2)} \cdot \frac{(x_2 - iy_2)}{(x_2 - iy_2)}$$

$$= \frac{x_1x_2 + iy_1x_2 - iy_2(x_1 + iy_1}{x^2 - (iy_2)^2}$$

$$= \frac{x_1x_2 + y_1y_2 + i(y_1x_2 - y_2x_1)}{x_2^2 + y_2^2}$$

$$= \frac{x_1x_2 + y_1y_2}{x_2^2 + y_2^2} + \frac{i(y_1x_2 - y_2x_1)}{x_2^2 + y_2^2} = a + ib$$

since $i^2 = -1$

Let $Z = a + ib$, equating real and imaginary terms, we have

$$a = \frac{x_1x_2 + y_1y_2}{x_2^2 + y_2^2} \quad \text{and} \quad b = \frac{y_1x_2 - y_2x_1}{x_2^2 + y_2^2}$$

Note that the quantity in the denominators after multiplying by the conjugate is always positive, $(x^2 + y^2)$.

WORKED EXAMPLE 8

Express $Z_1 = \frac{1 - i3}{2 + i5}$ and $Z_2 = \frac{3 - i4}{-3 - i4}$ in the form $a + ib$ and find Z_1Z_2 and $\frac{Z_1}{Z_2}$.

SOLUTION 8

$$Z_1 = \frac{1 - i3}{2 + i5} \times \frac{2 - i5}{2 - i5} = \frac{2 - i6 - i5 - 15}{4^2 + 5^2} = \frac{-13}{41} - \frac{i11}{41}$$

$$Z_1 = \frac{-13}{41} - \frac{i11}{41}$$

$$Z_2 = \frac{3 - i4}{-3 - i4} \times \frac{-3 + i4}{-3 + i4} = \frac{-9 + 12i + i12 + 16}{(-3)^2 + 4^2} = \frac{7}{25} + \frac{24}{25}i$$

$$Z_2 = \frac{7}{25} + \frac{i24}{25}$$

$$Z_1Z_2 = \left(\frac{-13}{41} - \frac{i11}{41}\right)\left(\frac{7}{25} + \frac{i24}{25}\right) = -\frac{91}{41 \times 25} - \frac{i77}{41 \times 25} - \frac{i13 \times 24}{41 \times 25} + \left(\frac{11 \times 24}{41 \times 25}\right)$$

$$= \frac{173}{1025} - i\frac{389}{1025}$$

$$\frac{Z_1}{Z_2} = \frac{-\frac{13}{41} - i\frac{11}{41}}{\frac{7}{25} + i\frac{24}{25}} \cdot \frac{\frac{7}{25} - i\frac{24}{25}}{\frac{7}{25} - i\frac{24}{25}} = \frac{\frac{-91}{1025} - \frac{i\,77}{1025} + \frac{i\,312}{1025} - \frac{264}{1025}}{\left(\frac{7}{25}\right)^2 + \left(\frac{24}{25}\right)^2}$$

$$= \frac{-355}{1025} + \frac{i235}{1025}$$

Simplifying we have

$$\frac{Z_1}{Z_2} = -\frac{71}{205} + \frac{i47}{205}$$

The real term of $\frac{Z_1}{Z_2}$ is $\frac{-71}{205}$ and the imaginary term of $\frac{Z_1}{Z_2}$ is $\frac{47}{205}$.

EXERCISES (5) & (6)

1. If a complex number $Z = x + iy$ and its conjugate $Z^* = x - iy$, show that (i) $ZZ^* = x^2 + y^2$ (ii) $\left(\frac{1}{Z}\right)^* = \frac{1}{Z^*}$.

2. Define the conjugate, Z^* of a complex number Z, and prove that if Z_1 and Z_2 are any complex numbers then $(Z_1 + Z_2)^* = Z_1^* + Z_2^*$.

3. If $\frac{1}{Z} = \frac{iy + x}{iy - x}$ prove that $\frac{x^2 + y^2}{y^2 - x^2} = \frac{2Z}{1 + Z^2}$

4. If $Z = x + iy$ where x and y are non zero numbers, find the cartesian equation in order that $\frac{Z}{1 + Z^2}$ is real. Express $\frac{Z}{1 + Z^2}$ in the form $a + ib$.

5. Given that $Z_1 = 1 + i$, $Z_2 = 1 - 2i$, $\frac{1}{Z} = \frac{1}{Z_1} + \frac{1}{Z_2}$ find Z in the form $a + ib$, where a and b are real.

6. Find the real numbers u and v given that $\frac{1}{u + iv} = 3 + 4i$

7. Find the real numbers x and y given that $\frac{1}{x + iy} = 5 - 12i$.

8. Express in the form $a + ib$,
$$\frac{3 + 4i}{5 - 12i}$$

9. Simplify
   (i) $(1 - i)^{-2} + (1 + i)^{-2}$
   (ii) $(1 + i)^{-3} + (1 - i)^{-3}$
   (iii) $(1 - i)^{-4} + (1 + i)^{-4}$

10. Find the real numbers x and y such that $(2 + i)x + (1 + 3i)y + 2 = 0$

11. Prove that (3,4) is one root of the equation $Z^2 - 6z + 25 = 0$ and find the other root.

12. If $Z_1 = a + b^2 - 3i$ and $Z_2 = 2 - ab^2 i$, determine the real values of a and b such that $Z_1 = \overline{Z}_2$ or $\overline{Z}_1 = Z_2$.

13. Determine the complex numbers which verify the equation
$$\overline{Z} = Z^2$$

14. If Z_1 and Z_2 are any two complex numbers, show that $Z_1\overline{Z}_2 + \overline{Z}_1 Z_2$ is real.

15. Determine the real numbers

    (i) $Z^2 + \overline{Z}^2$ (ii) $\frac{Z + 1}{\overline{Z}} + \frac{\overline{Z} + 1}{Z}$

7. DEFINES THE MODULUS AND ARGUMENT OF COMPLEX NUMBERS

Let $Z = x + iy$ be a complex number where x and y are real quantities. The modulus of Z is denoted as $|Z|$ due to the Weirstrass notation and means the magnitude of the vector quantity or sometimes is called the absolute value of the complex number.

$$|Z| = \sqrt{(x)^2 + (y)^2} = r.$$

The argument of Z is denoted by arg Z and means the angle of the vector quantity or sometimes is called "the amplitude of the complex number ".

The angle is measured with reference to the positive x-axis and in an anticlockwise direction

$$\arg Z = \tan^{-1} y/x = \theta.$$

It is necessary to illustrate the modulus and argument of a complex number in an Argand diagram and to use it when evaluating these quantities.

Fig. 8 illustrates this clearly.

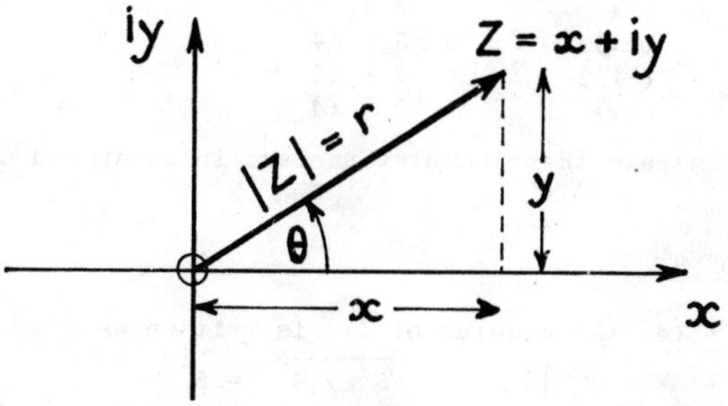

Fig.8 Modulus and argument of a complex number.

8. CONVERTS THE CARTESIAN FORM $x + iy$ INTO POLAR FORM $r(\cos\theta + i\sin\theta)$ and vice versa

$$Z = x + iy$$
$$|Z| = \sqrt{x^2 + y^2} = r$$
$$\arg Z = \tan^{-1} y/x = \theta$$

and from Fig.8 $\cos\theta = x/r$ and $\sin\theta = y/r$

$$Z = x + iy = r\cos\theta + \sin\theta\, r$$
$$Z = r(\cos\theta + i\sin\theta)$$

$\cos\theta + i\sin\theta$ may be abbreviated to $\underline{/\theta}$ or cis, the former is an engineer's notation and the latter that of a mathematician

$$Z = r\underline{/\theta} = r\, cis\, \theta = r(\cos\theta + i\sin\theta).$$

WORKED EXAMPLE 9

Find the moduli and arguments of the following complex numbers:

(i) $Z_1 = 3 + i4$
(ii) $Z_2 = -3 + i4$
(iii) $Z_3 = -3 - i4$
(iv) $Z_4 = 3 - i4$

and illustrate these complex numbers in an Argand diagram.

SOLUTION 9

$Z_1 = 3 + i4$ the modulus of Z_1 is written as

$$|Z_1| = \sqrt{3^2 + 4^2} = 5$$

The argument of Z_1 is the angle θ_1 since $\tan\theta_1 = 4/3$

$$\arg Z_1 = \theta_1 = \tan^{-1} 4/3 = 53^\circ\, 7'\, 48'' \simeq 53^\circ\, 8'$$
$$Z_1 = 5\underline{/53^\circ\, 8'} = 5(\cos 53^\circ\, 8' + i\sin 53^\circ\, 8').$$
$$Z_2 = -3 + i4$$
$$|Z_2| \quad \sqrt{(-3)^2 + (4)^2} = 5$$
$$\arg Z_2 = \theta_2 = 180^\circ - \tan^{-1} 4/3 = 180^\circ - 53^\circ\, 8'$$
$$= 126^\circ\, 52'.$$

Since $\tan\theta_2 = 4/-3$

$$Z_2 = 5\underline{/126^\circ\, 52'} = 5\cos 126^\circ\, 52' + i\sin 126^\circ\, 52').$$
$$Z_3 = -3 - i4$$
$$|Z_3| = \sqrt{(-3)^2 + (-4)^2} = 5$$
$$\arg Z_3 = \theta_3 = 180^\circ + \tan^{-1} 4/3$$
$$= 180^\circ + 53^\circ\, 8'$$

Since $\tan\theta_2 = -4/-3$

if we cancel the negative signs $\tan\theta_2 = 4/3$, which implies that $\theta_2 = 53^\circ\, 8$, which is not correct.

$$Z_3 = 5\underline{/233^\circ\ 8'} = 5(\cos 233^\circ\ 8' + i \sin 233^\circ\ 8'),$$

$$\text{or}\quad Z_3 = 5\underline{/-126^\circ\ 52'}$$

$$Z_4 = 3 - i4.$$

It is better therefore, to evaluate θ_1 and use the Argand diagram to find the exact angle:

$$|Z_4| = \sqrt{(3)^2 + (-4)^2} = 5$$

$$\arg Z_4 = \theta_4 = 360^\circ - \tan^{-1} \tfrac{4}{3} = 360^\circ - 53^\circ\ 8'$$

Since $\tan \theta_4 = -\tfrac{4}{3}$

$$Z_4 = 5\underline{/306^\circ\ 52'} = 5(\cos 306^\circ\ 52' + i \sin 306^\circ\ 52')$$

$$\text{or}\quad Z_4 = \underline{/-53^\circ\ 08'}.$$

The moduli are 5 and the arguments of the angles of the complex numbers are shown θ_1, θ_2, θ_3, θ_4, and are measured with Ox as a reference in an anticlockwise direction.

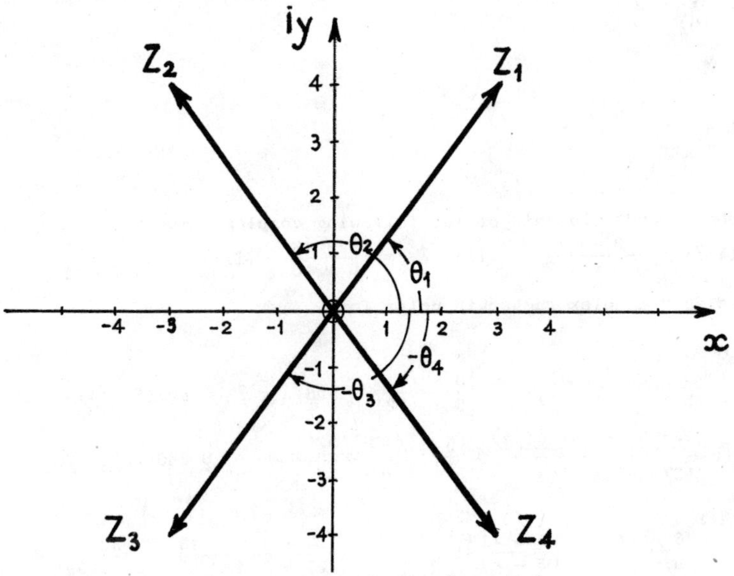

Fig.9 Moduli and Arguments of complex numbers. Principal values $-\pi \leq \theta \leq \pi$.

Fig. 9 shows the four complex numbers Z_1, Z_2, Z_3, and Z_4, and Fig.9(a), Fig.9(b), Fig.9(c) and Fig.9(d) show these complex numbers separately for simplicity.
Note that $\cos \theta + i \sin \theta$ is written as $\underline{/\theta}$ which is a very useful notation for abbreviation.

$\cos \theta - i \sin \theta$ is written as $\underline{/-\theta}$ or $\overline{\diagdown\theta}$

Remember $\cos(-\theta) = \cos \theta$ which is an even function
$\sin(-\theta) = -\sin \theta$ which is an odd function

therefore $\underline{/\theta}$ represents $\cos \theta + i \sin \theta$ in shorthand

and $\overline{\diagdown\theta}$ represents $\cos \theta - i \sin \theta$ in shorthand, or $\underline{/-\theta}$

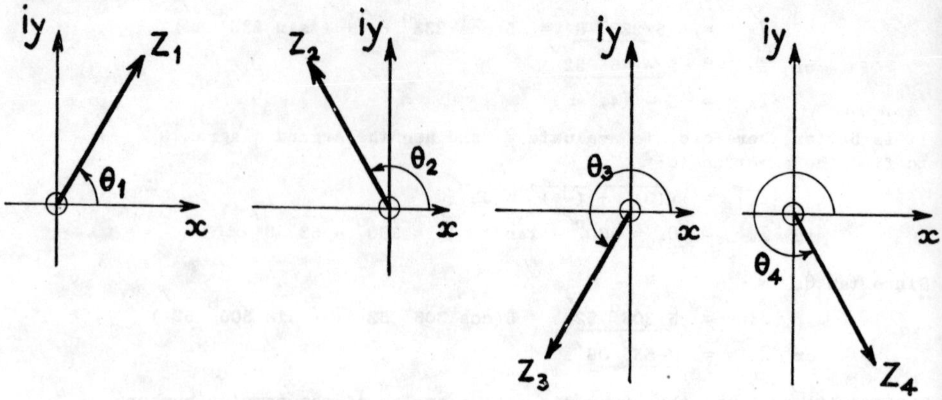

Fig.9 (a)) Moduli and Arguments of complex numbers
(b)) $(0^{\circ} \leqslant \theta \leqslant 360^{\circ})$
(c))
(d))

EXAMPLE 10

Find the moduli and arguments of the following complex numbers:

(i) $Z_1 = \dfrac{3 - i4}{5 + i12}$ (ii) $Z_2 = \dfrac{1 - i3}{2 + i5}$ (iii) $Z_3 = \dfrac{3 - i4}{-3 - i4}$

and express each complex number in polar form.

SOLUTION 10

(i) $Z_1 = \dfrac{3 - i4}{5 + i12}$, $|Z_1| = \dfrac{|3 - i4|}{|5 + i12|}$, and $\arg Z_1 = \arg(3 - i4) - \arg(5 + i12)$

$$Z_1 = \frac{5/360^{\circ} - \tan^{-1}(^4/_3)}{13/\tan^{-1}(^{12}/_5)} = \frac{5}{13}\,\frac{/306^{\circ}\ 52'}{/67^{\circ}\ 23'} = 0.385\ /239^{\circ}\ 29'$$

Alternatively

$$Z_1 = \frac{3 - i4}{5 + i2} \cdot \left(\frac{5 - i12}{5 - i12}\right) = \frac{15 - i20 - i36 - 48}{25 + 144} = \frac{-33 - i56}{169}$$

Multiplying numerator and denominator by the conjugate $(5 - i12)$, $Z_1 = \dfrac{-33}{169} - \dfrac{i56}{169}$

$$|Z_1| = \sqrt{\left(\frac{-33}{169}\right)^2 + \left(\frac{-56}{169}\right)^2} \quad \text{and } \arg Z_1 = 180 + \tan^{-1}\ {}^{56}/_{33}$$

$$|Z_1| = 0.385 \quad \text{and } \arg Z_1 = 239^{9}\ 29'$$

$$\boxed{Z_1 = 0.385\ /239^{\circ}\ 29'.}$$

It is observed that the alternative method is lengthier

(ii) $Z_2 = \dfrac{1 - i3}{2 + i5}$, $|Z_2| = \left|\dfrac{1 - i3}{2 + i5}\right| = \dfrac{\sqrt{(1)^2 + (-3)^2}}{\sqrt{(2)^2 + (5)^2}} = \dfrac{\sqrt{10}}{\sqrt{29}} = 0.587$

$$\arg\ Z_2 = \arg.\ \frac{1 - i3}{2 + i5} = \arg\ (1 - i3) - \arg\ (2 + i5)$$

$$= -\tan^{-1} 3 - \tan^{-1} 2.5 = -71^\circ\ 34' - 68^\circ\ 12'$$

$$= -139^\circ\ 46' \text{ or } +220^\circ\ 14'$$

$$\boxed{Z_2 = 0.587\ \underline{/-139^\circ\ 48'}} \text{ or } 0.587\ \overline{\backslash 139^\circ\ 46'}$$

$$Z_2 = 0.587\ (\cos 139^\circ\ 48' - i \sin 139^\circ\ 48')$$

(iii) $$Z_3 = \frac{3 - i4}{-3 - i4}, \quad |Z_3| = \frac{|3 - i4|}{|-3 - i4|} = \frac{\sqrt{(3)^2 + (-4)^2}}{\sqrt{(-3)^2 + (-4)^2}} = \frac{5}{5} = 1$$

$$\arg\ Z_3 = (3 - i4) - \arg\ (-3 - i4)$$

$$= -\tan^{-1} 4/3 - (180^\circ + 53^\circ\ 8') = -53^\circ\ 8' - 180^\circ - 53^\circ\ 8'$$

$$= -286^\circ\ 16' = 73^\circ\ 44'$$

$$\boxed{Z_3 = 1\underline{/\ 73^\circ\ 44'}} = (\cos 73^\circ\ 44' + i \sin 73^\circ\ 44').$$

There are two methods in finding the modulus of a quotient, such as

$$Z_1 = \frac{3 - i4}{5 + i12}$$

(a) Either we find the modulus of the numerator and divide by the modulus of the denominator,

(b) or by rationalising the expression by multiplying the numerator and denominator by the conjugate of the denominator.

The purpose of this is to obtain a real quantity in the denominator.

To find, however, the argument of Z_1 there are also two ways -

The first method is to find the arguments of the numerator and denominator of the individual complex numbers and subtract that of the denominator from the numerator.
This is proved later in Chapter 10.

The argument of Z_1 is $\theta_1 - \theta_2$, that is, the argument of $(3 - i4)$ minus the argument of $(5 + i12)$.

The second method, although more straightforward, results in tedious calculations.

9. MULTIPLIES AND DIVIDES COMPLEX NUMBERS USING THE POLAR FORM

The cartesian or quadratic form of complex numbers is useful in adding or subtracting complex numbers, where the real parts are either added or subtracted and the imaginary parts are either added or subtracted.

The polar form is extremely useful in multiplying and dividing complex numbers where the moduli are either multiplied or divided and their arguments are either added or subtracted.

$$Z_1 = r_1 \underline{/\theta_1} = r_1 (\cos\theta_1 + i \sin\theta_2)$$

$$Z_2 = r_2 \underline{/\theta_2} = r_2 (\cos\theta_2 + i \sin\theta_2)$$

$$Z_1 Z_2 = r_1 r_2 \underline{/\theta_1 + \theta_2} = r_1 r_2 (\cos\theta_1 + i \sin\theta_2)(\cos\theta_2 + i \sin\theta_2)$$

$$\frac{Z_1}{Z_2} = \frac{r_1}{r_2} \underline{/\theta_1 - \theta_2} = \frac{r_1 (\cos\theta_1 + i \sin\theta_2)}{r_2 (\cos\theta_2 + i \sin\theta_2)}$$

<u>WORKED EXAMPLE 11</u>

Multiply and divide the complex numbers

$$Z_1 = 3\underline{/35^\circ} \quad \text{and} \quad Z_2 = 5\underline{/-45^\circ}$$

<u>SOLUTION 11</u>

$$Z_1 . Z_2 = (3\underline{/35^\circ}) . (5\underline{/-45^\circ}) = 15\underline{/35^\circ - 45^\circ} = 15\underline{/-10^\circ} \text{ or } 15\underline{/350^\circ}$$

$$\boxed{Z_1 Z_2 = 15\underline{/350^\circ}}$$

$$\frac{Z_1}{Z_2} = \frac{3\underline{/35^\circ}}{5\underline{/-45^\circ}} = 0.6\underline{/35^\circ - (-45^\circ)} = 0.6\underline{/80^\circ}$$

$$\boxed{\frac{Z_1}{Z_2} = 0.6\underline{/80^\circ}}$$

GEOMETRIC REPRESENTATION OF COMPLEX NUMBERS

(a) THE SUM AND DIFFERENCE OF TWO COMPLEX NUMBERS

Let vectors $\vec{OP_1}$ and $\vec{OP_2}$ represent two complex numbers Z_1 and Z_2, as shown in Fig. 10

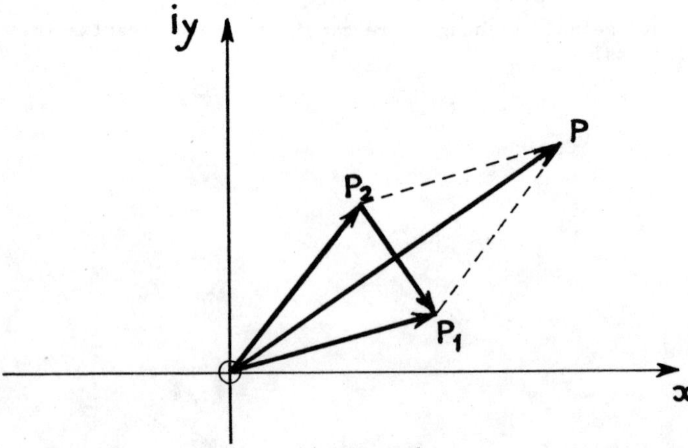

Fig.10 The sum and difference of two complex numbers.

To find the sum of the vectors $\vec{OP}_1$ and $\vec{OP}_2$, the parallelogram is constructed as shown, hence the resultant is the diagonal $\vec{OP} = Z_1 + Z_2 = \vec{OP}_1 + \vec{OP}_2$, which is the sum.

To find the difference of the vectors $\vec{OP}_1$ and $\vec{OP}_2$ draw P_2P_1, then from the triangle OP_2P_1,

$$\vec{OP}_2 + \vec{P_2P}_1 = \vec{OP}_1 \quad \text{and}$$
$$\vec{P_2P}_1 = \vec{OP}_1 - \vec{OP}_2 = Z_1 - Z_2$$

the difference of the vectors

(b) THE PRODUCT OF TWO COMPLEX NUMBERS

To find geometrically the product of two vectors or two complex numbers.

Two similar triangles OAP_1 and OP_2P are formed where the angles $O\hat{P}_2P$ and $O\hat{A}P_1$ are equal, and $P\hat{O}P_2$ and $P_1\hat{O}A$ are equal, hence $O\hat{P}P_2 = O\hat{P}_1A$.

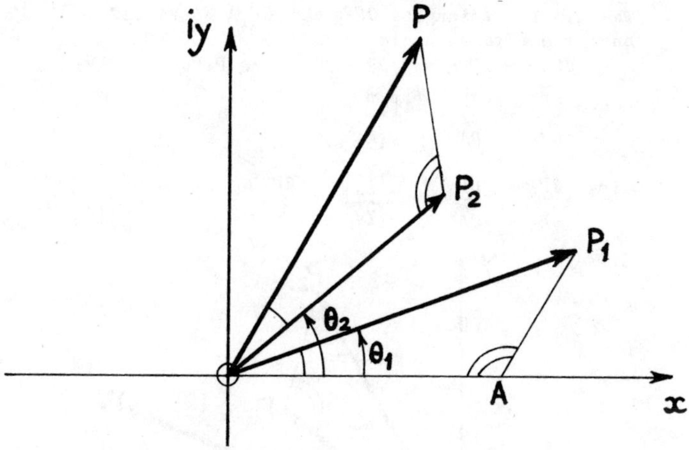

Fig.11 The product of two complex numbers
$OP = Z_1Z_2 \quad OP_2 = Z_2 \quad OP_1 = Z_1$
OP_2P, OAP_1 are similar triangles.

From the similar triangles where $OA = 1$

$$\frac{\vec{OA}}{\vec{OP}_2} = \frac{\vec{OP}_1}{\vec{OP}} = \frac{\vec{AP}_1}{\vec{PP}_2}$$

$$\frac{1}{|Z_2|} = \frac{|Z_1|}{\vec{OP}} = \frac{AP_1}{\vec{PP}_2}$$

From $\dfrac{1}{|Z_2|} = \dfrac{|Z_1|}{\vec{OP}} \quad \rightleftarrows \quad \vec{OP} = |Z_1||Z_2| = |Z_1Z_2|$

where $\vec{OP}_1 = |Z_1|$ the magnitude of Z_1
$\vec{OP}_2 = |Z_2|$ the magnitude of Z_2

$$|Z_1Z_2| = |Z_1||Z_2|$$

$$\text{Let } X\hat{O}P_1 = \theta_1 = \text{argument of } Z_1 = \arg Z_1$$
$$X\hat{O}P_2 = \theta_2 = \text{argument of } Z_2 = \arg Z_2$$
$$X\hat{O}P = \theta = \text{argument of } Z_1Z_2 = \arg Z_1Z_2$$

$$X\hat{O}P = X\hat{O}P_2 + P_2\hat{O}P = X\hat{O}P_2 + \theta_1$$
$$\theta = \theta_2 + P_2\hat{O}P = \theta_2 + \theta_1$$
$$P_2\hat{O}P = \theta - \theta_2 = \theta_1$$
$$\theta = \arg Z_1 + \arg Z_2 = \arg (Z_1Z_2)$$

$$\arg (Z_1Z_2) = \arg Z_1 + \arg Z_2$$

(c) <u>THE QUOTIENT OF TWO COMPLEX NUMBERS</u>

To find geometrically the quotient of two vectors or complex numbers:

The similar triangles OPP_2 and OP_1A of Fig.12 have their three angles equal,

$$O\hat{P}_2P = O\hat{P}_1A \quad P_2\hat{O}P = P_1\hat{O}A \text{ and } P_2\hat{P}O = O\hat{A}P_1$$

$$\text{hence } \frac{\vec{OP_2}}{\vec{OP_1}} = \frac{\vec{OP}}{\vec{OA}} = \frac{P_2P}{P_1A}$$

$$\text{using } \frac{\vec{OP_2}}{\vec{OP_1}} = \frac{\vec{OP}}{\vec{OA}} \qquad \frac{|Z_2|}{|Z_1|} = \frac{\vec{OP}}{1} \qquad \vec{OP} = \frac{|Z_2|}{|Z_1|}$$

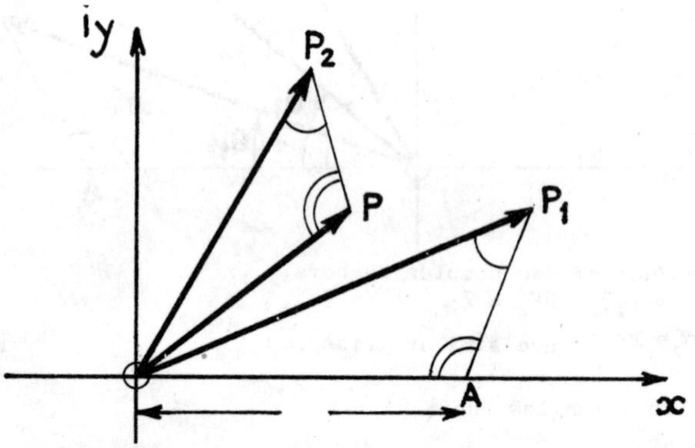

Fig.12 The quotient of two complex numbers
$OP_2 = Z_2 \quad OP_1 = Z_1 \quad OP = Z_2/Z_1.$

OP_2P and OP_1A **are similar triangles.**

$$\frac{OP}{1} = \frac{OP_2}{OP_1} = \frac{Z_2}{Z_1}$$

To show that $\arg \dfrac{Z_2}{Z_1} = \arg Z_2 - \arg Z_1$

$$\text{Let } X\hat{O}P = \theta_1$$
$$X\hat{O}P = \theta_2$$
$$X\hat{O}P = \theta$$
$$X\hat{O}P = X\hat{O}P_2 - P_2\hat{O}P = X\hat{O}P_2 - X\hat{O}P_1$$
$$\theta = \theta_2 - \theta_1$$
$$\arg \frac{Z_2}{Z_1} = \theta = \arg Z_2 - \arg Z_1$$

EXERCISES (7) (8) (9)

1. **Calculate the modulus and the argument of the complex numbers (principal values) :**

(i) 1
(ii) -1
(iii) $-i$
(iv) $1 + i\sqrt{3}$
(v) $1 - i\sqrt{3}$
(vi) $-1 + i\sqrt{3}$
(vii) $-1 - i\sqrt{3}$
(viii) $i - \sqrt{3}$
(ix) $i + 1$
(x) $-1 + i$
(xi) $-1 - i$
(xii) $1 - i$
(xiii) $i - 1$
(xiv) $\sqrt{3} - i$
(xv) $-i - \sqrt{3}$
(xvi) $2 + i3$
(xvii) $-3 + i4$
(xviii) $-2 - 4i$
(xix) $3 - 2i$
(xx) $5 - i3$

Sketch these complex numbers in an Argand diagram, and express them in polar form.

2. **If** $(x + iy)^n = a + bi$, **express** $a^2 + b^2$ **in terms of** x, y, **and the argument of** $a + bi$ **in terms of** n, x **and** y.

3. **Express in polar form the following:**

(i) $3 + 4i$
(ii) $3 - 4i$
(iii) $-3 + 4i$
(iv) $-3 - 4i$
(v) $\sqrt{2} - i$
(vi) $\sqrt{3} - i$
(vii) $\cos\alpha - i\sin\alpha$
(viii) $\sin\alpha + i\cos\alpha$
(ix) $\sin\alpha - i\cos\alpha$
(x) $\cos\alpha + i\sin\alpha$
(xi) $1 + i\tan\alpha$
(xii) $1 + i\cot\alpha$
(xiii) $\tan\beta - i$
(xiv) $\cos\alpha - i\sin\beta + i(\sin\alpha + i\cos\beta)$
(xv) $1 + r\cos\phi + ir\sin\phi$.

4. **Express in quadratic form the following complex numbers:**

(i) $1\underline{/0^{\circ}}$
(ii) $3\underline{/-30^{\circ}}$
(iii) $1\underline{/\pi/4}$
(iv) $5\underline{/-\pi/2}$
(v) $3\underline{/\pi}$
(vi) $\underline{/-180^{\circ}}$
(vii) $3(\cos\theta + i\sin\theta)$
(viii) $1\underline{/353^{\circ}}$
(ix) $3\underline{/360^{\circ}}$
(x) $7\underline{/4\pi/3}$

5. Express $Z = \dfrac{1 + 2i}{3 + 4i}$ in the form $x + iy$ where x and y are real, and hence calculate the modulus and argument of Z.

6. A complex number Z has a modulus $\sqrt{2}$ and an argument of $\pi/3$. Write down this complex number in

   (a) quadratic or cartesian form
   (b) polar form
   (c) exponential form.

7. If $Z = 3 + i4$, find $\dfrac{1}{Z}$, Z^2 and Z^3 and plot these values on an Argand diagram.

8. Mark in an Argand diagram the points P_1 and P_2 which represent the two complex numbers $Z_1 = -1 - i$ and $Z_2 = 1 + i\sqrt{3}$. On the same diagram, mark the points P_3 and P_4 which represent $(Z_1 - Z_2)$ and $(Z_1 + Z_2)$ respectively.

   Find the modulus and argument of

   (i) Z_1 (ii) Z_2 (iii) Z_1Z_2 (iv) $\dfrac{Z_1}{Z_2}$ (v) $\dfrac{Z_2}{Z_1}$

9. If $Z = \cos\theta + i(1 + \sin\theta)$, show that the value of the modulus of $\dfrac{2Z - i}{iZ - 1}$ is unity.

10. If $Z = 1 + i3$ and $Z_2 = 3 - i$ show in an Argand diagram points representing the complex numbers

$$Z_1,\ Z_2,\ Z_1Z_2,\ Z_1 + Z_2,\ \frac{Z_2}{Z_1},\ \frac{Z_1}{Z_2}.$$

10 DEFINES THE EXPONENTIAL FORM OF A COMPLEX NUMBER

Applying $e^x = 1 + \frac{x}{1!} + \frac{x}{2!} + \frac{x}{3!} + \ldots\ldots$

$$e^{i\theta} = 1 + i\theta + \frac{(i\theta)^2}{2!} + \frac{(i\theta)^3}{3!} + \frac{(i\theta)^4}{4!} + \ldots\ldots$$

$$= 1 + i\theta - \frac{\theta^2}{2!} - \frac{i\theta^3}{3!} + \frac{\theta^4}{4!} - \ldots\ldots$$

$$= \left(1 - \frac{\theta^2}{2!} + \frac{\theta^4}{4!} - \ldots\right) + i\left(\theta - \frac{\theta^3}{3!} + \frac{\theta^5}{5!} - \ldots\right)$$

$$\boxed{e^{i\theta} = \cos\theta + i\sin\theta}$$ **EULER'S FORMULA**

therefore,

$$Z = r(\cos\theta + i\sin\theta) = re^{i\theta} \quad r\exp i\theta$$

$$\boxed{Z = re^{i\theta}}$$

The exponential form of a complex number where θ <u>is expressed in radians</u>.

It is easily now seen that the product of two numbers

$Z_1 = 3e^{i\pi/6}$ and $Z_2 = 5e^{i\pi/4}$

is $Z_1 Z_2 = 15e^{i\pi/4 + i\pi/6} = 15e^{i5\pi/12}$

and the quotient of these two numbers is $\frac{Z_1}{Z_2} = \frac{3e^{i\pi/6}}{5e^{i\pi/4}} = 0.6e^{-i\pi/12}$ by applying the law of indices.

It is evident that the exponential form of a complex number is useful in dividing and multiplying complex numbers.
It is therefore important for the student to be familiar with all the forms of the complex numbers and exercise in manipulating all the forms $Z = x + iy = r(\cos\theta + i\sin\theta) = re^{i\theta} = r\underline{/\theta}$.

WORKED EXAMPLE 12

Find the product of the complex numbers $Z_1 = 1\underline{/\pi/4}$ and $Z_2 = 2\underline{/\pi/3}$ and the quotients $\frac{Z_1}{Z_2}$ and $\frac{Z_2}{Z_1}$.

SOLUTION 12

$Z_1 Z_2 = 1\underline{/\pi/4} \,.\, 2\underline{/\pi/3} = 2\underline{/\pi/4 + \pi/3} = 2\underline{/7\pi/12}$

$$\boxed{Z_1 Z_2 = 2\underline{/7\pi/12}}$$

$$\frac{Z_1}{Z_2} = \frac{1}{2}\frac{\underline{/\pi/4}}{\underline{/\pi/3}} = 0.5\underline{/\pi/4 - \pi/3} = 0.5\underline{/\pi/12}$$

$$\frac{Z_2}{Z_1} = \frac{2}{1}\underline{/\pi/3 - \pi/4} = 2\underline{/-\pi/12}$$

$$\boxed{\frac{Z_2}{Z_1} = 2\underline{/-\pi/12}}$$

EXERCISES (10)

1. Express the following complex numbers in the cartesian and polar forms:

(i)	$Z_1 = 3e^{-i\pi/2}$	(vi)	$Z_6 = 4e^{i5\pi/6}$
(ii)	$Z_2 = 5e^{i\pi}$	(vii)	$Z_7 = -3e^{-i11\pi/6}$
(iii)	$Z_3 = e^{\pi/4\,i}$	(viii)	$Z_8 = e^{-i\pi/4}$
(iv)	$Z_4 = e^{-\pi/3\,i}$	(ix)	$Z_9 = e^{-i3}$
(v)	$Z_5 = e^{i3\pi/2}$	(x)	$Z_{10} = e^{-i}$.

2. Express the following complex numbers in the exponential form:

(i)	$Z_1 = 0 - 3i$	(vi)	$Z_6 = -2\sqrt{3} + i2$
(ii)	$Z_2 = -5 + 0i$	(vii)	$Z_7 = \frac{-3\sqrt{3}}{2} - \frac{i3}{2}$
(iii)	$Z_3 = \frac{1}{\sqrt{2}} + i\frac{1}{\sqrt{2}}$	(viii)	$Z_8 = \frac{1}{\sqrt{2}} - i\frac{1}{\sqrt{2}}$
(iv)	$Z_4 = \frac{1}{2} - i\frac{\sqrt{3}}{2}$	(ix)	$Z_9 = -0.99 - i0.14$
(v)	$Z_5 = 0 - i$	(x)	$Z_{10} = 0.54 - i0.84$.

3. Express the following complex numbers in the exponential form:

(i)	$Z_1 = 3\underline{/-\pi/2}$	(vi)	$Z_6 = 4\underline{/5\pi/6}$
(ii)	$Z_2 = 3\underline{/\pi}$	(vii)	$Z_7 = -3\underline{/-11\pi/6}$
(iii)	$Z_3 = 1\underline{/\pi/4}$	(viii)	$Z_8 = 1\underline{/-\pi/4}$
(iv)	$Z_4 = 1\underline{/-\pi/3}$	(ix)	$Z_9 = 1\underline{/171^\circ}\ 53'$
(v)	$Z_5 = 1\underline{/3\pi/2}$	(x)	$Z_{10} = 1\underline{/-57^\circ}\ 18'$.

4. Show that $\cos(\theta + \pi/4)e^{i\pi/4} - \sin(\theta - \pi/4)e^{-i\pi/4} = (\cos\theta - \sin\theta)$.

5. If $Z = \cos\frac{\pi}{2} + i\sin\frac{\pi}{2}$ find the value of Z^2 and deduce the value of Z^3.

6. If $Z_1Z_2 = 3 + i4$ and $\frac{Z_1}{Z_2} = 5i$, and the arguments of Z_1 and Z_2 lie between $-\pi$ and $+\pi$.

Determine the complex numbers Z_1 and Z_2 in
(i) cartesian form
(ii) polar form
(iii) exponential form.

11 DETERMINES THE SQUARE ROOTS OF A COMPLEX NUMBER.

Find the square root of $x + iy$, i.e $\sqrt{x + iy}$.

Let $\sqrt{x + iy} = a + ib$

Squaring up both sides:

$$x + iy = (a + ib)^2 = a^2 + 2abi + i^2b^2$$
$$x + iy = (a^2 - b^2) + i2ab$$

Equating real terms:

$$a^2 - b^2 = x \quad \ldots\ldots\ldots \quad (1)$$

Equating the imaginary terms:

$$2ab = y$$
$$(a^2 + b^2)^2 = (a^2 - b^2)^2 + 4a^2b^2 = x^2 + y^2$$
$$a^2 + b^2 = \sqrt{x^2 + y^2} \quad \ldots\ldots\ldots (2)$$

Adding equations (1) and (2)

$$2a^2 = \sqrt{x^2 + y^2} + x$$

Subtracting equation (1) from (2)

$$2b^2 = \sqrt{x^2 + y^2} - x$$

Therefore the required real values a and b, are given

$$a = \sqrt{\frac{\sqrt{x^2 + y^2} + x}{2}} \qquad b = \sqrt{\frac{\sqrt{x^2 + y^2} - x}{2}}$$

WORKED EXAMPLE 13

Find the square roots of $3 + i4$

SOLUTION 13

Let $\sqrt{3 + i4} = \pm(a + ib)$

Squaring up both sides:

$$3 + i4 = a^2 + i^2b^2 + i\,2ab$$
$$= (a^2 - b^2) + i\,2ab$$

Equating real and imaginary terms:

$$a^2 - b^2 = 3$$
$$2ab = 4$$

From $a = \sqrt{\frac{\sqrt{x^2 + y^2} + x}{2}}$ and $b = \sqrt{\frac{\sqrt{x^2 + y^2} - x}{2}}$

$$a = \sqrt{\frac{\sqrt{x^2 + y^2} + x}{2}} = \sqrt{\frac{5 + 3}{2}} = 2$$

where $\sqrt{x^2 + y^2} = \sqrt{3^2 + 4^2} = \sqrt{9 + 16} = \sqrt{25} = 5$

and $b = \sqrt{\frac{\sqrt{x^2 + y^2} - x}{2}} \quad \sqrt{\frac{5 - 3}{2}} = 1$

therefore $a = 2$ and $b = 1$

$$\sqrt{3 + i4} = \pm(2 + i).$$

WORKED EXAMPLE 14

Verify that $3 - i7$ is one of the square roots of $-40 - 42i$.
Write down the other square root.

SOLUTION 14

$3 - i7 = \sqrt{-40 - 42i}$

Squaring up both sides of this equation $(3 - i7)^2 = -40 - 42i$.
The L.H.S. $(3 - i7)^2$ is given as $(3 - i7)^2 = 9 - 42i + i^2 49 = -40 - 42i$.

Therefore $3 - i7$ is one of the square roots of $-40 - 42i$.
The other square root will be $-(-40 - 42i) = 40 + 42i$
since $(3 - i7)^2 = -40 - 42i$
and $3 - i7 = \pm(-40 - 42i)^{1/2}$

One root is $-40 - 42i$ and the other $40 + 42i$.

EXERCISES 11

1. Verify that $(1 - i)$ is one of the square roots of $0 - 2i$.
Write down the other square root.

2. Verify that $3 + i4$ is one of the square roots of $-7 + 24i$.
Write down the other square root.

3. Verify that $7 - i12$ is one of the square roots of $-95 - 168i$.
Write down the other square root.

4. Verify that i is one of the square roots of -1.
Write down the other square root.

5. Verify that $-3 - i4$ is one of the square roots of $-7 + 24i$.
Write down the other square root.

6. Determine the square roots of the following complex numbers:

(i)	$7 - i$	(vi)	$1 + 3i$
(ii)	$-1 + i$	(vii)	$4 + 7i$
(iii)	$-3 - i4$	(viii)	$-1 + 3i$
(iv)	$4 - i3$	(ix)	$-4 - i4$
(v)	$-3 + 7i$	(x)	$-6 + i$

7. Verify that $4 - i4$ is one of the square roots of $-32i$.
Write down the other square root.

8. If $\pm(a + ib)$ are the square roots of $3 - i4$.
Find a and b.

12. PROOF OF DE MOIVRE THEOREM BY INDUCTION AND OTHERWISE

De Moivre, Abraham, an English mathematician was born on May 26th 1667 at Vitry Champagne and died on November 27th 1754, in London. He was of French extraction.

De Moivre became famous as a mathematician and was elected F.R.S. in 1697. His contributions to trigonometry are two well known theorems concerning expansions of trigonometrical functions.

De Moivre's theorem states: $(\cos\theta + i\sin\theta)^n = \cos n\theta + i\sin n\theta$.

<u>Proof by Induction</u> where n is an integer

For $n = 1$

$$(\cos\theta + i\sin\theta)^1 = \cos\theta + i\sin\theta \quad \text{which is true.}$$

For $n = k$

$$(\cos\theta + i\sin\theta)^k = \cos k\theta + i\sin k\theta \quad \text{it would be true}$$

For $n = k + 1$,

$$\begin{aligned}(\cos\theta + i\sin\theta)^{k+1} &= (\cos k\theta + i\sin k\theta)(\cos\theta + i\sin\theta)\\ &= \cos k\theta\cos\theta + i\sin k\theta\cos\theta\\ &\quad + i\sin\theta\cos k\theta - \sin k\theta\sin\theta\\ &= \cos(k\theta + \theta) + i\sin(k\theta + \theta)\\ &= \cos(k+1)\theta + i\sin\theta(k+1)\end{aligned}$$

therefore it is true for n.

<u>Proof of De Moivre's Theorem</u>

If n is any rational number

$\cos n\theta + i\sin n\theta$ is a value of $(\cos\theta + i\sin\theta)$

(i) If n is a positive integer

let $Z_1 = r(\cos\theta + i\sin\theta) \equiv (r\cos\theta, r\sin\theta)$

$Z_2 = s(\cos\phi + i\sin\phi) \equiv (s\cos\phi, s\sin\phi)$

$$\begin{aligned}&(r\cos\theta, r\sin\theta) \times (s\cos\phi, s\sin\phi)\\ &= \{rs(\cos\theta\cos\phi - \sin\theta\sin\phi), rs(\sin\theta\cos\phi + \cos\theta\sin\phi)\}\\ &= \{rs\cos(\theta + \phi), rs\sin(\theta + \phi)\}\\ &= rs\{\cos(\theta + \phi) + i\sin(\theta + \phi)\}\end{aligned}$$

$$|Z_1 Z_2| = |Z_1| \,.\, |Z_2|$$

$$\arg Z_1Z_2 = \arg Z_1 + \arg Z_2$$

$$r_1 r_2 r_3 \ldots\ldots r_n(\cos\theta_1 + i\sin\theta_1)(\cos\theta_2 + i\sin\theta_2)(\cos\theta_3 + i\sin\theta_3)\ldots$$

$= r^n\{\cos(\theta_1 + \theta_2 + \ldots) + i\sin(\theta_1 + \theta_2 + \ldots)\}$

if $r_1 = r_2 = \ldots = r_n = r$

$= r^n(\cos n\theta + i\sin n\theta)$ if $\theta_1 = \theta_2 = \ldots\ldots = \theta_n = \theta$

Therefore $(\cos\theta + i\sin\theta)^n = \cos n\theta + i\sin n\theta$

(ii) If n is a negative integer
put $n = -m$

$$(\cos\theta + i\sin\theta)^n = (\cos\theta + i\sin\theta)^{-m} = \frac{1}{(\cos\theta + i\sin\theta)^m}$$
$$= \frac{1}{(\cos m\theta + i\sin m\theta)}$$

but $(\cos m\theta + i\sin m\theta)(\cos m\theta - i\sin m\theta) = \cos^2 m\theta - i^2\sin^2 m\theta$

therefore
$$\frac{1}{(\cos m\theta + i\sin m\theta)} = \cos m\theta - i\sin m\theta = \cos(-m\theta) + i\sin(m\theta)$$
$$(\cos\theta + i\sin\theta)^n = \cos n\theta + i\sin n\theta$$

Therefore, if n is a positive or negative integer, there is only one value of $(\cos\theta + i\sin\theta)^n$, and this value is $\cos n\theta + i\sin n\theta$.

(iii) If n is a fraction, i.e. put $n = p/q$ where p, q are integers and q is positive.

In this case, $(\cos\theta + i\sin\theta)^n$ has many values.

To show that there are q values:

Let $\left(\cos\frac{p}{q}\theta + i\sin\frac{p}{q}\theta\right)^q = \cos p\theta + i\sin p\theta$

since $(\cos n\theta + i\sin n\theta)$ is a value of $(\cos\theta + i\sin\theta)^n$.

Also $\cos p\theta + i\sin p\theta = (\cos\theta + i\sin\theta)^p$

since $(\cos n\theta + i\sin n\theta)$ is a value of $(\cos\theta + i\sin\theta)^n$.

Also $\cos p\theta + i\sin p\theta = (\cos\theta + i\sin\theta)^p$.

Therefore, $\left(\cos\frac{p}{q}\theta + i\sin\frac{p}{q}\theta\right)^q = (\cos\theta + i\sin\theta)^p$.

It follows that $\cos\frac{p\theta}{q} + i\sin\frac{p\theta}{q}$ is a value of $(\cos\theta + i\sin\theta)^{p/q}$ by the definition of a^n.

The theorem is therefore proved for all rational values of

$(\cos\theta + i\sin\theta)^{p/q} = \left[(\cos\theta + i\sin\theta)^p\right]^{1/q}$

and has therefore q distinct roots

$$(\cos\theta + i\sin\theta)^{p/q} = (\cos p\theta + i\sin p\theta)^{1/q}$$
$$= \cos\left(\frac{p}{q}\theta + 2k\pi\right) + i\sin\left(\frac{p}{q}\theta + 2k\pi\right)$$

where $k = 0, 1, 2, \ldots\ldots (q-1)$.

The principal value of $(\cos\theta + i\sin\theta)^{p/q}$ is taken to be $\cos\frac{p}{q}\theta + i\sin\frac{p}{q}\theta$, only if $-\pi \le \theta \le \pi$

THE PRINCIPAL ROOT

The root whose vector is nearest to the positive x-axis, is called 'the principal root'.

The cube roots of unity are $Z_1 = \underline{/0^o}$, $Z_2 = \underline{/2\pi/3}$, $Z_3 = \underline{/4\pi/3}$

The principal root is Z_1, which is the nearest to the positive x-axis.

13 **Expands cos $n\theta$, sin $n\theta$ and tan $n\theta$, where n is any positive integer.**

$\cos n\theta + i \sin n\theta = (\cos\theta + i \sin\theta)^n = (c + is)^n$

where $c = \cos\theta$, $s = \sin\theta$,

$$(c + is)^n = c^n + nc^{n-1}is + n(n-1)c^{n-2} i^2 s^2 \frac{1}{2!} + n(n-1)(n-2)c^{n-3}i^3 s^3 \frac{1}{3!} + \ldots + i^n s^n$$

$$= \{c^n - n(n-1)c^{n-2} \frac{s^2}{2!} + \ldots\} + i\{nc^{n-1} s - n(n-1)(n-2)c^{n-3} \frac{s^3}{3!} \ldots\}$$

Equating real and imaginary terms, we have

$$\cos n\theta = c^n - \frac{n(n-1)}{2} c^{n-2} s^2 + n(n-1)(n-2)(n-3)\, c^{n-4} \frac{s^4}{4!} - \ldots.$$

the real terms,

$$\sin n\theta = nc^{n-1}s - n(n-1)(n-2)\, c^{n-3} s^3 \frac{1}{3!} + \ldots\ldots$$

the imaginary terms.

$\cos n\theta + i \sin n\theta = \cos^n\theta\, (1 + i \tan\theta)^n = \cos^n\theta\, (1 + it)^n$

where $t = \tan\theta$

$$\cos n\theta = \cos^n\theta\{1 - {}^nC_2\, t^2 + {}^nC_4\, t^4 - \ldots\ldots\}$$

where ${}^nC_r = \dfrac{n!}{(n-r)!\, r!}$

$$\sin n\theta = \cos^n\theta\, \{{}^nC_1\, t - {}^nC_3\, t^3 + \ldots\ldots\}$$

$$\tan n\theta = \frac{\{{}^nC_1\, t - {}^nC_3\, t^3 + \ldots\ldots\}\cos^n\theta}{\{1 - {}^nC_2\, t^2 + {}^nC_4\, t^4 - \ldots\}\cos^n\theta} = \frac{{}^nC_1\, t - {}^nC_3\, t^3 + \ldots\ldots}{1 - {}^nC_2\, t^2 + {}^nC_4\, t^4 - \ldots.}$$

WORKED EXAMPLE 15

(a) Determine an expression for tan 3θ in terms of tan θ.

(b) State De Moivre's theorem as an integral exponent, and write down the value $(\cos\theta + i \sin\theta)^5$ as a multiple angle.

(c) Simplify $\dfrac{(\cos\theta_1 + i \sin\theta_1)^3}{(i \cos\theta_2 + \sin\theta_2)^4}$

SOLUTION 15

(a) $\cos 3\theta + i \sin 3\theta = (\cos\theta + i \sin\theta)^3$

$$= \cos^3\theta + 3\cos^2\theta.\, i \sin\theta + 3\cos\theta\, i^2\sin^2\theta + i^3\sin^3\theta$$

$$= (\cos^3\theta - 3\cos\theta\sin^2\theta) + i\,(3\cos^2\theta\sin\theta - \sin^3\theta)$$

Equating real and imaginary terms

$$\cos 3\theta = \cos^3\theta - 3\cos\theta\sin^2\theta$$

$$\sin 3\theta = 3\cos^2\theta\sin\theta - \sin^3\theta$$

$$\tan 3\theta = \frac{\sin 3\theta}{\cos 3\theta} = \frac{3\cos^2\theta\sin\theta - \sin^3\theta}{\cos^3\theta - 3\cos\theta\sin^2\theta}$$

$$= \frac{\cos^3\theta\left(\dfrac{3\sin\theta}{\cos\theta} - \dfrac{\sin^3\theta}{\cos^3\theta}\right)}{\cos^3\theta\left(1 - 3\dfrac{\sin^2\theta}{\cos^2\theta}\right)} = \frac{3\tan\theta - \tan^3\theta}{1 - 3\tan^2\theta}$$

Therefore

$$\boxed{\tan 3\theta = \frac{3\tan\theta - \tan^3\theta}{1 - 3\tan^2\theta}}$$

(b) $(\cos\theta + i\sin\theta)^n = \cos n\theta + i\sin n\theta$ De Moivre's theorem.

$(\cos\theta + i\sin\theta)^5 = \cos 5\theta + i\sin 5\theta$

(c) $$\frac{(\cos\theta_1 + i\sin\theta_1)^3}{(i\cos\theta_2 + \sin\theta_2)^4} = \frac{(\cos\theta_1 + i\sin\theta_1)^3}{\{i(\cos\theta_2 - i\sin\theta_2)\}^4}$$

$$= \frac{(\cos\theta_1 + i\sin\theta_1)^3}{i^4(\cos\theta_2 - i\sin\theta_2)^4}$$

$$= \frac{\cos 3\theta_1 + i\sin 3\theta_1}{\cos 4\theta_2 - i\sin 4\theta_2} = (\cos 3\theta_1 + i\sin 3\theta_1).(\cos 4\theta_2 + i\sin 4\theta_2)$$

$$= \{\cos(3\theta_1 + 4\theta_2) + i\sin(3\theta_1 + 4\theta_2)\}$$

14. APPLICATION OF DE MOIVRE'S THEOREM.

To express $\sin\theta$, $\cos\theta$, $\sin n\theta$ and $\cos n\theta$ in terms of Z where $Z = \cos\theta + i\sin\theta$.

Let $Z = \cos\theta + i\sin\theta$

$$\frac{1}{Z} = (\cos\theta + i\sin\theta)^{-1} = \cos(-\theta) + i\sin(-\theta) = \cos\theta - i\sin\theta$$

$$Z - \frac{1}{Z} = \cos\theta + i\sin\theta - (\cos\theta - i\sin\theta)$$
$$= 2i\sin\theta$$

$$\sin\theta = \frac{1}{2i}\left(Z - \frac{1}{Z}\right)$$

$$Z + \frac{1}{Z} = \cos\theta + i\sin\theta + \cos\theta - i\sin\theta = 2\cos\theta$$

$$\cos\theta = \frac{1}{2}\left(Z + \frac{1}{Z}\right)$$

$$Z^n = (\cos\theta + i\sin\theta)^n = \cos n\theta + i\sin n\theta$$

$$\frac{1}{Z^n} = (\cos\theta - i\sin\theta)^n = \cos n\theta - i\sin n\theta$$

$$Z^n - \frac{1}{Z^n} = \cos n\theta + i\sin n\theta - (\cos n\theta - i\sin n\theta)$$
$$= 2i\sin n\theta$$

$$\sin n\theta = \frac{1}{2i}\left(Z^n - \frac{1}{Z^n}\right)$$

$$Z^n + \frac{1}{Z^n} = \cos n\theta + i\sin n\theta + \cos n\theta - i\sin n\theta$$
$$= 2\cos n\theta$$

$$\cos n\theta = \frac{1}{2}\left(Z^n + \frac{1}{Z^n}\right).$$

WORKED EXAMPLE 16

Expand $\left(Z + \frac{1}{Z}\right)^3$ and $\left(Z - \frac{1}{Z}\right)^3$ where $Z = \cos\theta + i\sin\theta$ and find the expression for $\cos^3\theta + \sin^3\theta$.

SOLUTION 16

Since $(a + b)^3 = a^3 + 3a^2b + 3ab^2 + b^3$, substitute $a = Z$ and $b = \frac{1}{Z}$

$$\left(Z + \frac{1}{Z}\right)^3 = Z^3 + 3Z^2 . \frac{1}{Z} + 3.Z.\left(\frac{1}{Z}\right)^2 + \frac{1}{Z^3}$$

$$= \left(Z^3 + \frac{1}{Z^3}\right) + 3\left(Z + \frac{1}{Z}\right) = 2\cos 3\theta + 3(2\cos\theta)$$

$$= 2\cos 3\theta + 6\cos\theta$$

$$\left(Z - \frac{1}{Z}\right)^3 = Z^3 - 3Z^2\frac{1}{Z} + 3Z . \frac{1}{Z^2} - \frac{1}{Z^3} = \left(Z^3 - \frac{1}{Z^3}\right) - 3\left(Z - \frac{1}{Z}\right)$$

$$= 2i\sin 3\theta - 6i\sin\theta$$

$$\cos^3\theta + \sin^3\theta = \frac{1}{2^3}\left(Z + \frac{1}{Z}\right)^3 + \frac{1}{2^3 i^3}\left(Z - \frac{1}{Z}\right)^3$$

$$= \frac{1}{8}(2\cos 3\theta + 6\cos\theta) + \frac{1}{8i^3}(2i\sin 3\theta - 6i\sin\theta)$$

$$= \frac{1}{4}\cos 3\theta + \frac{3}{4}\cos\theta - \frac{1}{4}\sin 3\theta + \frac{3}{4}\sin\theta$$

$$= \frac{1}{4}(\cos 3\theta - \sin 3\theta) + \frac{3}{4}(\cos\theta + \sin\theta)$$

WORKED EXAMPLE 17

Express (a) $\sin 3\theta$ in terms of $\sin\theta$ and
(b) $\cos 3\theta$ in terms of $\cos\theta$

SOLUTION 17

$(\cos\theta + i\sin\theta)^3 = \cos 3\theta + i\sin 3\theta$
using De Moivre's theorem.

Also $(\cos\theta + i\sin\theta)^3$ can be expanded using Binomial theorem.

$$(\cos\theta + i\sin\theta)^3 = \cos^3\theta + 3\cos^2\theta\, i\sin\theta$$

$$+ \frac{3 \times 2}{1 \times 2}\cos\theta\, i^2\sin^2\theta + \frac{3 \times 2 \times 1}{1 \times 2 \times 3} i^3\sin^3\theta$$ using Binomial expansion

$$\cos 3\theta + i\sin 3\theta = \cos^3\theta + 3i\cos^2\theta\sin\theta - 3\sin^2\theta\cos\theta - i\sin^3\theta$$

Equating real and imaginary terms

$$\cos 3\theta = \cos^3\theta - 3\cos\theta\sin^2\theta = \cos^3\theta - 3\cos\theta(1 - \cos^2\theta)$$

$$\sin 3\theta = 3\cos^2\theta\sin\theta - \sin^3\theta = 3(1 - \sin^2\theta).\sin\theta - \sin^3\theta$$

$$\boxed{\sin 3\theta = 3\sin\theta - 4\sin^3\theta}$$

$$\boxed{\cos 3\theta = 4\cos^3\theta - 3\cos\theta}$$

WORKED EXAMPLE 18

Express (a) $\sin 5\theta$ in terms of $\sin\theta$ and
(b) $\cos 5\theta$ in terms of $\cos\theta$

SOLUTION 18

$(\cos\theta + i\sin\theta)^5 = \cos 5\theta + i\sin 5\theta$ by De Moivre's Theorem

Expanding by the binomial theorem

$$(\cos\theta + i\sin\theta)^5 = \cos^5\theta + 5\cos^4\theta\, i\sin\theta + \frac{5 \times 4}{1 \times 2}\cos^3\theta\, i^2\sin^2\theta$$

$$+ \frac{5 \times 4 \times 3}{1 \times 2 \times 3}\cos^2\theta\, i^3\sin^3\theta + \frac{5 \times 4 \times 3 \times 2}{1 \times 2 \times 3 \times 4}\cos\theta\sin^4\theta$$

$$+ \frac{5 \times 4 \times 3 \times 2 \times 1}{1 \times 2 \times 3 \times 4 \times 5}\, i^5\sin^5\theta$$

$$\cos 5\theta + i\sin 5\theta = \cos^5\theta + i5\cos^4\theta\sin\theta - 10\cos^3\theta\sin^2\theta - 10i\cos^2\theta\sin^3\theta + 5\cos\theta\sin^4\theta + i\sin^5\theta$$

Equate real and imaginary terms

$$\cos 5\theta = \cos^5\theta - 10\cos^3\theta\sin^2\theta + 5\cos\theta\sin^4\theta$$

$$= \cos^5\theta - 10\cos^3\theta(1 - \cos^2\theta) + 5\cos\theta(1 - \cos^2\theta)^2$$

$$= \cos^5\theta - 10\cos^3\theta + 10\cos^5\theta + 5\cos\theta - 10\cos^3\theta + 5\cos^5\theta$$

$$\boxed{\cos 5\theta = 16\cos^5\theta - 20\cos^3\theta + 5\cos\theta}$$

$$\sin 5\theta = 5\cos^4\theta\sin\theta - 10\cos^2\theta\sin^3\theta + \sin^5\theta$$

$$= 5(1 - \sin^2\theta)^2\sin\theta - 10(1 - \sin^2\theta)\sin^3\theta + \sin^5\theta$$

$$= (5 - 10\sin^2\theta + 5\sin^4\theta)\sin\theta - 10\sin^3\theta + 10\sin^5\theta + \sin^5\theta$$

$$= 5\sin\theta - 10\sin^3\theta + 5\sin^5\theta - 10\sin^3\theta + 10\sin^5\theta + \sin^5\theta$$

$$\sin 5\theta = 5\sin\theta - 10\sin^3\theta - 10\sin^3\theta + 10\sin^5\theta + 6\sin^5\theta$$

$$\boxed{\sin 5\theta = 5\sin\theta - 20\sin^3\theta + 16\sin^5\theta}$$

WORKED EXAMPLE 19

Using De Moivre's theorem, show that

$\sin 4\theta = 4\cos^3\theta \sin\theta - 4\cos\theta \sin^3\theta$ and

$\cos 4\theta = \cos^4\theta - 6\cos^2\theta \sin^2\theta + \sin^4\theta$

hence $\tan 4\theta = \dfrac{4t - 4t^3}{1 - 6t^2 + t^4}$

where $t = \tan\theta$.

Hence find the values of $\tan {}^{\pi}/_8$ and $\tan {}^{3\pi}/_8$ in surd forms.

SOLUTION 19

$$(\cos\theta + i\sin\theta)^4 = \cos 4\theta + i\sin 4\theta$$

$$= \cos^4\theta + 4\cos^3\theta\, i\sin\theta + \frac{4 \times 3}{1 \times 2}\cos^2\theta\, i^2 \sin^2\theta$$

$$+ \frac{4 \times 3 \times 2}{1 \times 2 \times 3}\cos\theta\, i^3 \sin^3\theta + \frac{4 \times 3 \times 2 \times 1}{1 \times 2 \times 3 \times 4} i^4 \sin^4\theta$$

Equating real and imaginary terms

$$\cos 4\theta = \cos^4\theta - 6\cos^2\theta\sin^2\theta + \sin^4\theta$$

$$\sin 4\theta = 4\cos^3\theta\sin\theta - 4\cos\theta\sin^3\theta$$

$$\tan 4\theta = \frac{4\cos^3\theta\sin\theta - 4\cos\theta\sin^3\theta}{\cos^4\theta - 6\cos^2\theta\sin^2\theta + \sin^4\theta}$$

Dividing numerator and denominator by $\cos^4\theta$

$$\tan 4\theta = \frac{4\tan\theta - 4\tan^3\theta}{1 - 6\tan^2\theta + \tan^4\theta} = \frac{4t - 4t^3}{1 - 6t^2 + t^4}$$

Let $\theta = \pi/8$

$$\tan 4(\pi/8) = \frac{4t - 4t^3}{1 - 6t^2 + t^4} = \text{ where } \tan \pi/2 = \infty$$

Therefore the denominator must be zero

$\therefore\ 1 - 6t^2 + t^4 = 0$ or $t^4 - 6t^2 + 1 = 0$

$$t^2 = \frac{6 \pm \sqrt{36 - 4}}{2} = \frac{6 \pm \sqrt{32}}{2} = 3 \pm 2\sqrt{2}$$

$\therefore\ t = \pm\sqrt{3 \pm 2\sqrt{2}}$

Therefore, there are four solutions for t.

There are four solutions for t if $\theta = 3\pi/8$.
The negative solutions are omitted since $\tan \pi/8$, $\tan 3\pi/8$ are positive.
Also $\tan \pi/8 < \tan \pi/4 = 1$ and $\tan 3\pi/8 > \tan \pi/4 = 1$

$t = \sqrt{3 - 2\sqrt{2}} = \sqrt{a} - \sqrt{b}$, squaring up both sides $3 - 2\sqrt{2} = a + b - 2\sqrt{ab}$
where $a = 2$, $b = 1$.

Therefore $\tan \pi/8 = \sqrt{2} - 1$

$t = \sqrt{3 + 2\sqrt{2}} = \sqrt{a} + \sqrt{b}$, squaring up both sides -

$3 + 2\sqrt{2} = a + b + 2\sqrt{ab}$ where $a = 2$, $b = 1$

Therefore $\tan 3\pi/8 = \sqrt{2} + 1$

EXERCISES (12) (13) (14)

1. Simplify (i) $\dfrac{\cos\phi - i\sin\phi}{\cos 2\phi + i\sin 2\phi}$

   (ii) $(\cos\theta - i\sin\theta)^7$

   (iii) $\dfrac{(\cos 2\theta - i\sin 2\theta)^4}{(\cos 3\theta + i\sin 3\theta)^3}$

   (iv) $(1 + \cos\theta + i\sin\theta)^3$

2. If $Z = \cos\theta + i\sin\theta$ express in terms of θ

   (i) $Z + \dfrac{1}{Z}$ (ii) $Z - \dfrac{1}{Z}$ (iii) $Z^n + \dfrac{1}{Z^n}$ (iv) $Z^n - \dfrac{1}{Z^n}$

3. Write down the square roots of

   (i) $\cos 2\theta - i\sin 2\theta$
   (ii) $\cos 3\theta + i\sin 3\theta$
   (iii) $\sin\theta + i\cos\theta$
   (iv) $-i$
   (v) i

4. If $Z = \cos\theta + i\sin\theta$, express $\sqrt{\dfrac{1+Z}{1-Z}}$ in the form $a + bi$

   (i) for $0 < \theta < \pi$
   (ii) $\pi < \theta < 2\pi$

5. Express $\cos^3\theta$, $\sin^3\theta$, $\cos^4\theta$, $\sin^4\theta$, $\cos^5\theta$, $\sin^5\theta$ in terms of multiple angles.

6. Write down the cube roots of

   (i) $\cos 3\theta - i\sin 3\theta$
   (ii) $-i$
   (iii) $\sin\theta - i\cos\theta$

7. Write down the roots of

   (i) $Z^5 = 1$
   (ii) $Z^4 - 1 = 0$

8. Simplify $(\cos\theta + i\sin\theta)^n + (\cos\theta + i\sin\theta)^{-n}$

9. Express $\sin 3\theta$, $\sin 4\theta$, $\sin 5\theta$, and $\cos 3\theta$, $\cos 4\theta$, $\cos 5\theta$, in terms of single angles.

10. Simplify $(\cos A + i\sin A)(\cos B + i\sin B)(\cos C + i\sin C)$ if $A + B + C = \pi$.

11. Find, in the form $a + ib$, the three roots of the equation
    $Z^3 - 7 + 24i = 0$

12. Express the square roots of $-2i$ in the form $\pm(a + ib)$, where a and b are real numbers.

13. Find the roots of the complex equations:

    (i) $Z^3 - 1 = 0$ (iii) $Z^3 + i = 0$
    (ii) $Z^3 - i = 0$ (iv) $Z^3 + 1 = 0$

RELATES HYPERBOLIC AND TRIGONOMETRIC FUNCTIONS

$e^{i\theta} = \cos\theta + i\sin\theta \qquad e^{-i\theta} = \cos\theta - i\sin\theta$

By definition of $\cosh x = \dfrac{e^x + e^{-x}}{2}$, we have that

$$\cosh i\theta = \frac{e^{i\theta} + e^{-i\theta}}{2} = \frac{\cos\theta + i\sin\theta + \cos\theta - i\sin\theta}{2}$$

$$\boxed{\cosh(i\theta) = \cos(\theta)} \text{..........................(1)}$$

By definition of $\sinh x = \dfrac{e^x - e^{-x}}{2}$ we have that

$$\sinh i\theta = \frac{e^{i\theta} - e^{-i\theta}}{2} = \frac{\cos\theta + i\sin\theta - \cos\theta + i\sin\theta}{2}$$

$$\boxed{\sinh(i\theta) = i\sin(\theta)} \text{....................(2)}$$

$$\sinh\theta = \frac{e^{\theta} - e^{-\theta}}{2} = \frac{1}{2}\left[1 + \theta + \frac{\theta^2}{2!} + \frac{\theta^3}{3!} + \ldots -(1 - \theta + \frac{\theta^2}{2!} - \frac{\theta^3}{3!} + \ldots)\right]$$

$$\sinh\theta = \theta + \frac{\theta^3}{3!} + \frac{\theta^5}{5!} + \ldots.$$

From the expansion $\sin x = x - \dfrac{x^3}{3!} + \dfrac{x^5}{5!}$ ...... we have that

$$\sin i\theta = (i\theta) - \frac{(i\theta)^3}{3!} + \frac{(i\theta)^5}{5!} - \ldots..$$

$$\sin i\theta = i\theta + \frac{i\theta^3}{3!} + \frac{i\theta^5}{5!} + \ldots = i\left(\theta + \frac{\theta^3}{3!} + \frac{\theta^5}{5!} + \ldots.\right)$$

Multiplying both sides by $-i$, we have $-i\sin i\theta = \theta + \dfrac{\theta^3}{3!} + \dfrac{\theta^5}{5!} + \ldots.$

$$\boxed{\sinh(\theta) = -i\sin(i\theta)} \text{.....................(3)}$$

Multiplying both sides by i, we have $\sin i\theta = i\sinh\theta$

$$\cosh\theta = \frac{e^{\theta} + e^{-\theta}}{2} = 1 + \frac{\theta^2}{2!} + \frac{\theta^4}{4!} + \ldots.$$

From the expansions, $\cos x = 1 - \dfrac{x^2}{2!} + \dfrac{x^4}{4!}$ ......, we have

$$\cos i\theta = 1 - \frac{(i\theta)^2}{2!} + \frac{(i\theta)^4}{4!} \ldots.$$

$$= 1 + \frac{\theta^2}{2!} + \frac{\theta^4}{4!} + \ldots..$$

The right hand side of this equation is $\cosh\theta$, therefore

$$\boxed{\cosh(\theta) = \cos(i\theta)} \quad \text{....................(4)}$$

Similarly we can show the circular functions to hyperbolic functions

$\sin\theta = i\sinh i$

$\cos\theta = \cosh i\theta$

$\sin i\theta = i\sinh\theta$

$\cos i\theta = \cosh\theta$

$\cos\theta + i\sin\theta = \cosh i\theta + \sinh i\theta$

CIRCULAR FUNCTIONS TO HYPERBOLIC FUNCTIONS

It is required to show that $\sin\theta = -i \sinh i\theta$.

The expansion of the series of $\sinh i\theta$

$$\sinh i\theta = (i\theta) + \frac{(i\theta)^3}{3!} + \frac{(i\theta)^5}{5!} + \ldots = i\theta + \frac{i^3\theta^3}{3!} + \frac{i^5\theta^5}{5!} + \ldots$$

$$= i\theta - \frac{i\theta^3}{3!} + \frac{i\theta^5}{5!} - \ldots$$

multiplying both sides by $-i$

$$-i \sinh i\theta = -i^2\theta + \frac{i^2\theta^3}{3!} - \frac{i^2\theta^5}{5!} +$$

$$= \theta - \frac{\theta^3}{3!} + \frac{\theta^5}{5!} - \ldots$$

The right hand is the expansion of $\sin\theta$

therefore $\boxed{\sin(\theta) = -i \sinh(i\theta)}$ ....(5)

It is required to show that $\cos\theta = \cosh i\theta$

The expansion of the series of $\cosh i\theta$:

$$\cosh i\theta = 1 + \frac{(i\theta)^2}{2!} + \frac{(i\theta)^4}{4!} + \ldots$$

$$= 1 + \frac{i^2\theta^2}{2!} + \frac{i^4\theta^4}{4!} + \ldots = 1 - \frac{\theta^2}{2!} + \frac{\theta^4}{4!} -$$

but $\cos\theta = 1 - \frac{\theta^2}{2!} + \frac{\theta^4}{4!} - \ldots$

therefore $\boxed{\cos(\theta) = \cosh(i\theta)}$ ....(6)

It is required to show that $\sin i\theta = i \sinh\theta$

The expansion of $\sin i\theta = (i\theta) - \frac{(i\theta)^3}{3!} + \frac{(i\theta)^5}{5!} - \ldots$

$$= i\theta - \frac{i^3\theta^3}{3!} + \frac{i^5\theta^5}{5!} - = i\theta + \frac{i\theta^3}{3!} + \frac{i\theta^5}{5!} + \ldots$$

and the expansion of $\sinh\theta = \theta + \frac{\theta^3}{3!} + \frac{\theta^5}{5!} + \ldots$

multiplying both sides by i, then

$\boxed{\sin(i\theta) = i \sinh(\theta)}$ ....(7)

The last expression to show is $\cos i\theta = \cosh\theta$

$$\cos i\theta = 1 - \frac{(i\theta)^2}{2!} + \frac{(i\theta)^4}{4!} - \ldots = 1 - \frac{i^2\theta^2}{2!} + \frac{i^4\theta^4}{4!} - \ldots$$

$$= 1 + \frac{\theta^2}{2!} + \frac{\theta^4}{4!} + \ldots$$

therefore $\boxed{\cos(i\theta) = \cosh(\theta)}$ ....(8)

WORKED EXAMPLE 20

Show that $\sin(x + iy) = \sin x \cosh y + i \cos x \sinh y$

SOLUTION 20

Using the addition theorem

$$\sin(x + y) = \sin x \cos y + \sin y \cos x$$

we have

$$\sin(x + iy) = \sin x \cos iy + \sin iy \cos x$$

but $\cos iy = \cosh y$ and $\sin iy = i \sinh y$,

then $\sin(x + iy) = \sin x \cosh y + i \cos x \sinh y$

WORKED EXAMPLE 21

Evaluate (i) $\sin(1 + i)^2$

(ii) $\cos(1 - i)^4$

(iii) $\sin(1 - i)^3$

in the form $p + id$

SOLUTION 21

(i) $\sin(1 + i)^2 = \sin(1 + i^2 + 2i) = \sin 2i = 2 \sin i \cos i$

$= 2i \sinh 1 \cosh 1 = 2i\,(1.175)\,(1.543) = i3.63$

since $\sin i = i \sinh 1$ and $\cos i = \cosh 1$

therefore $\sin(1 + i)^2 = i3.63$ which is purely imaginary

(ii) $\cos(1 - i)^4 = \cos\left(1 - 4i + \frac{4 \times 3}{1 \times 2}(-\ )^2 + \frac{4 \times 3 \times 2}{1 \times 2 \times 3}(-i)^3 + (-i)^4\right)$

$= \cos(1 - 4i - 6 + 4i + 1)$

$= \cos(-4) = \cos 4 = -0.654$

therefore $\cos(1 - i)^4 = -0.654$ which is purely real

(iii) $\sin(1 - i)^3 = \sin(1 - 3i + 3i^2 - i^3) = \sin(1 - 3i - 3 + i)$

$= \sin(-2 - 2i) = -\sin(2 + 2i)$

$= -\sin 2 \cos 2i - \sin 2i \cos 2$

$= -\sin 2 \cosh 2 - i \sinh 2 \cos 2$

$= -0.909\,(3.762) - i\,(3.627)\,(-0.416)$

$= -3.4197 + i1.51$

$= -3.42 + i1.51$

WORKED EXAMPLE 22

Find an expression for $\tan(x + iy)$ and evaluate $\tan(3 + i4)$.

...

SOLUTION 22

$$\tan(x+iy) = \frac{\sin(x+iy)}{\cos(x+iy)} = \frac{\tan x + \tan iy}{1 - \tan x \tan iy}$$

$$= \frac{\tan x + \dfrac{\sin iy}{\cos iy}}{1 - \tan x . \dfrac{\sin iy}{\cos iy}} = \frac{\tan x + \dfrac{i \sinh y}{\cosh y}}{1 - \tan x . \dfrac{i \sinh y}{\cosh y}}$$

$$= \frac{\tan x + i \tanh y}{1 - i \tan x \tanh y}$$

$$\tan(3+i4) = \frac{\tan 3 + i \tanh 4}{1 - i \tan 3 \tanh 4} = \frac{-0.1425 + i\,0.999}{1 - i(0.1425)(0.999)}$$

$$= \frac{-0.1425 + i\,0.999}{1 - i\,0.142} \times \frac{1 + i\,0.142}{1 + i\,0.142}$$

$$= \frac{-0.1425 + i\,0.999 - i\,0.0202 - 0.1418}{1 + 0.142^2}$$

$$= \frac{-0.284 + i\,0.9788}{1.020164}$$

$$= -0.278 + i\,0.959$$

$$\tan(3+i4) = -0.28 + i\,0.96$$

16 THE LOGARITHM OF A NEGATIVE NUMBER

Let $\log_e (-1) = Z$

$e^Z = -1$ by the definition of a logarithm

$= \cos \pi + i \sin \pi = e^{i\pi}$

then $Z = i\pi$

$$\boxed{\log_e (-1) = i\pi}$$

$$\log_e (-3) = Z$$

By definition $e^Z = -3 = 3(-1) = 3e^{i\pi}$

Taking logarithms on both sides:

$\log_e (-3) = \log_e (3)(-1) = \log_e 3 + \log_e (-1) = \ln 3 + i\pi$

$\log_e (-3) = \log_e 3 + i\pi \quad = 1.099 + i3.14159$

The logarithm of a negative number is a complex number which may be expressed in quadratic, polar or exponential.

$\ln N = \ln |N| + i\pi$ where N is a negative number

$\ln N = \sqrt{[\ln |N|]^2 + \pi^2} \quad \underline{/ \tan^{-1} \frac{\pi}{|N|}}$

$= \sqrt{[\ln |N|]^2 + \pi^2} \quad e^{i \tan^{-1} \pi/|N|}$

The logarithm of a vector $\ln(re^{i\theta}) = \ln r + \ln e^{i\theta}$

$= \ln r + i\theta$

WORKED EXAMPLE 23

Determine (i) $\log_e i$ (ii) $\log_e (1 - i)$ (iii) $\log_e (-1 + i)$

....

SOLUTION 23

(i) To find $\log_e i$

i can be written as $1\underline{/\pi/2}$

$\log_e i = \log_e 1 \underline{/\pi/2} = \ln 1 + \ln e^{i\pi/2} = i\pi/2$

(ii) $\log_e (1 - i) = \ln \sqrt{2} e^{-i\pi/4} = \ln \sqrt{2} + \ln e^{-i\pi/4}$

$= \frac{1}{2} \ln 2 + (-i\pi/4) = \frac{1}{2} \ln 2 - i\pi/4$

$|1 - i| = \sqrt{1 + (-1)^2} = \sqrt{2}$

$\arg (1 - i) = -\pi/4$

(iii) $\ln (-1 + i) = \ln \sqrt{2}\, e^{i\, 3\pi/4}$

$= \ln \sqrt{2} + \ln e^{i3\pi/4}$

$= \frac{1}{2} \ln 2 + i\, 3\pi/4$

WORKED EXAMPLE 24

Express $\ln \dfrac{3 - i4}{1 + i2}$ in the form $p + id$.

....

SOLUTION 24

Let $W = \dfrac{3 - i4}{1 + i2} = \dfrac{3 - i4}{1 + i2} \times \dfrac{1 - i2}{1 - i2} = \dfrac{3 - i4 - 6i - 8}{1 + 4} = \dfrac{-5}{5} - \dfrac{i10}{5}$

$W = -1 - i2$

$|W| = \sqrt{(-1)^2 + (-2)^2} = \sqrt{5}$

$\arg W = \pi + \tan^{-1} 2$

$\ln \dfrac{3 - i4}{1 + i2} = \ln \sqrt{5}.\ e^{i(\pi + \tan^{-1} 2)} = \ln \sqrt{5} + i\ (\pi + \tan^{-1} 2)$

$= \frac{1}{2} \ln 5 + i4.24874 = 0.81 + i4.25$

$\ln \dfrac{3 - i4}{1 + i2} = 0.81 + i4.25$

WORKED EXAMPLE 25

Find the principal value of i^i.

...

SOLUTION 25

Let $W = i^i$

Taking logs on both sides to the base e

$\log_e W = i \log_e i = i \ln 1 \underline{/\ \pi/2} = i \ln e^{i\pi/2}$

$= i \left(i \frac{\pi}{2}\right) = -\frac{\pi}{2}$

therefore $\boxed{e^{-\pi/2} = W}$

WORKED EXAMPLE 26

Evaluate 3^i correct to three decimal places.

....

SOLUTION 26

Let $Z = 3^i$

$\ln Z = i \ln 3 = i\, 1.099$

By definition $e^{i1.099} = Z = \cos 1.099 + i \sin 1.099$

$Z = 0.454 + i0.891$

$3^i = 0.454 + i0.891$

17 THE ROOTS OF EQUATIONS

Determines the cube roots of unity.

To find the roots of the cubic equation $Z^3 - 1 = 0$,
$(Z - 1)(Z^2 + Z + 1) = 0$, where $Z - 1 = 0$ or $Z^2 + Z + 1 = 0$
hence $Z = 1$ and

$$Z = \frac{-1 \pm \sqrt{1-4}}{2} = \frac{-1 \pm \sqrt{-3}}{2} = -\frac{1}{2} \pm i\frac{\sqrt{3}}{2}$$

$$Z_1 = 1, \quad Z_2 = -\frac{1}{2} + i\frac{\sqrt{3}}{2} \text{ and } Z_3 = -\frac{1}{2} - i\frac{\sqrt{3}}{2}$$

If $w = -\frac{1}{2} - \frac{\sqrt{3}}{2}i$ then $w^2 = -\frac{1}{2} + \frac{\sqrt{3}}{2}i$

The roots of the cubic equation are $1, w, w^2$

$$1 + w + w^2 = 1 - \frac{1}{2} - \frac{\sqrt{3}}{2}i - \frac{1}{2} + \frac{\sqrt{3}}{2}i = 0$$

$$\boxed{1 + w + w^2 = 0}$$

Alternatively $Z^3 - 1 = 0$

$$Z^3 = 1$$
$$Z = 1^{1/3}$$

The cube roots of unity are found as follows:

$$Z = \left(1\underline{/0^\circ}\right)^{1/3}$$

Remember $\underline{/\theta} \equiv \cos\theta + i\sin\theta$, and since the power is rational, we add $0^\circ + 2k\pi = 2k\pi$

$$Z = 1\left(\underline{/2k\pi}\right)^{1/3} = \underline{/2k\pi/3} \quad \text{using De Moivre's theorem}$$

where $k = 0, 1, 2$

then $Z_1 = 1\underline{/0^\circ}$, $Z_2 = 1\underline{/2\pi/3}$, $Z_3 = 1\underline{/4\pi/3}$

or $Z_1 = 1$, $Z_2 = \cos 2\pi/3 + i \sin 2\pi/3 = -\frac{1}{2} + i\frac{\sqrt{3}}{2}$

and $Z_3 = \cos 4\pi/3 + i\sin 4\pi/3 = -\frac{1}{2} - i\frac{\sqrt{3}}{2}$

The cube roots of unity are $1, -\frac{1}{2} + i\frac{\sqrt{3}}{2}, -\frac{1}{2} - i\frac{\sqrt{3}}{2}$ as before.

Fig. 13 represents these roots in an Argand diagram.
The two roots appear as a conjugate pair.

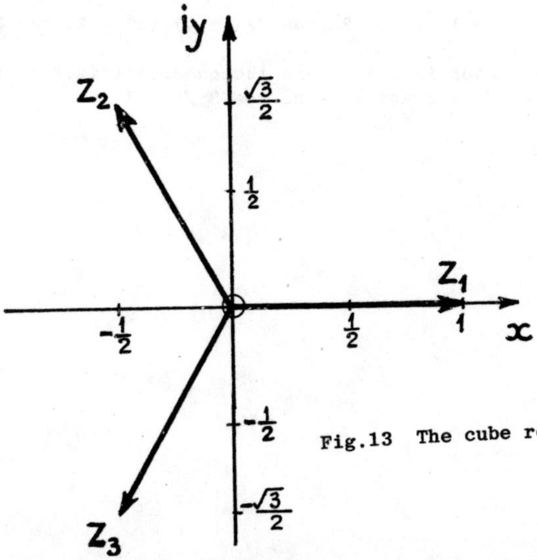

Fig.13 The cube roots of unity.

WORKED EXAMPLE 27

Factorize (a) $Z^5 - 32 = 0$) in a linear and
(b) $Z^5 + 1 = 0$) two quadratic factors.

SOLUTIONS 27

(a) $Z^5 - 32 = 0$ $\quad Z = 2(1)^{1/5} = 2(\cos 0 + i \sin 0)^{1/5}$

$Z^5 = 32$ $\quad Z = 2\left(\cos \frac{2k\pi}{5} + i \sin \frac{2k\pi}{5}\right)$

$Z = 32^{1/5}$

where $k = 0, \pm 1, \pm 2$

$Z_1 = 2$

$Z_2 = \underline{2/2\pi/_5} = 2(\cos 2\pi/_5 + i \sin 2\pi/_5)$

$Z_3 = \underline{2/-2\pi/_5} = 2(\cos 2\pi/_5 - i \sin 2\pi/_5)$

$Z_4 = \underline{2/4\pi/_5} = 2(\cos 4\pi/_5 + i \sin 4\pi/_5)$

$Z_5 = \underline{2/-4\pi/_5} = 2(\cos 4\pi/_5 - i \sin 4\pi/_5)$

The factors therefore are as follows:

$$(Z - 2)(Z - 2\cos 2\pi/_5 - 2i \sin 2\pi/_5)(Z - 2\cos 2\pi/_5 + 2i \sin 2\pi/_5)$$

$$= (Z - 2\cos 4\pi/_5 - 2i \sin 4\pi/_5)(Z - 2\cos 4\pi/_5 + i2 \sin 4\pi/_5)$$

$$= (Z - 2)\left[(Z - 2\cos 2\pi/_5)^2 + 4\sin^2 2\pi/_5\right]$$

$$\left[(Z - 2\cos 4\pi/_5)^2 + 4\sin^2 4\pi/_5\right]$$

$$= (Z - 2)\ (Z^2 - 4Z\cos 2\pi/_5 + 4)\ (Z^2 - 4Z\cos 4\pi/_5 + 4)$$

The linear factor is $(Z - 2)$ and the quadratic factors are $(Z^2 - 4Z \cos 2\pi/_5 + 4)$ and $(Z^2 - 4Z \cos 4\pi/_5 + 4)$.

(b) $Z^5 + 1 = 0$

$Z^5 = -1 = (\cos \pi + i \sin \pi)$

$Z = (\cos \pi + i \sin \pi)^{1/5} = \left(\cos \frac{\pi + 2k\pi}{5} + i \sin \frac{\pi + 2k\pi}{5}\right)$

where $k = 0, \pm 1, \pm 2$

$Z_1 = \underline{/\pi/_5}$, $Z_2 = \underline{/3\pi/_5}$, $Z_3 = \underline{/-\pi/_5}$, $Z_4 = \underline{/\pi}$, $Z_5 = \underline{/-3\pi/_5}$

$$(Z - \cos \pi/_5 - i \sin \pi/_5)(Z - \cos 3\pi/_5 - i \sin 3\pi/_5)$$

$$(Z - \cos \pi/_5 + i \sin \pi/_5)(Z + 1)(Z - \cos 3\pi/_5 + i \sin 3\pi/_5)$$

$$= (Z + 1)\ (Z^2 - 2Z\cos \pi/_5 + 1)\ (Z^2 - 2Z\cos 3\pi/_5 + 1)$$

The linear factor is $Z + 1$, and the quadratic factors are $Z^2 - 2Z \cos \pi/_5 + 1$ and $Z^2 - 2Z \cos 3\pi/_5 + 1$

WORKED EXAMPLE 28

Find the five roots of the $Z^5 + i = 0$, and plot them on an Argand diagram.

SOLUTION 28

$Z^5 + i = 0$
$Z^5 = -i$
$Z = (-i)^{1/5}$

To express $-i$ in the form $\cos\theta + i\sin\theta$, represent it in an Argand diagram. Fig.14

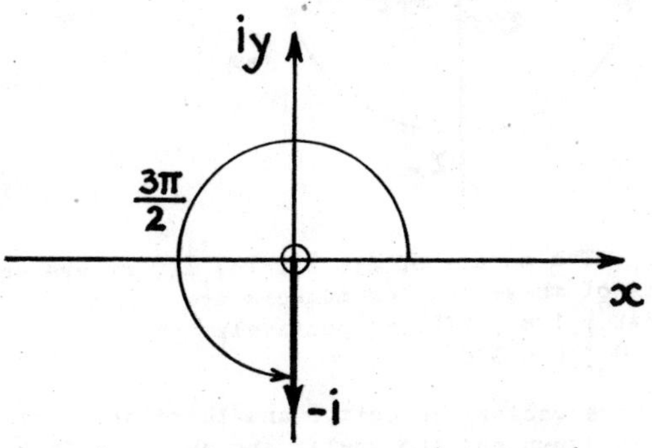

Fig.14 To express $-i$ in the form $\cos\theta + i\sin\theta$ $-i$ $= \cos 3\pi/2 + i\sin 3\pi/2$.

$Z = (\underline{/\ 3\pi/2})^{1/5} = \left(\underline{/(3\pi/2 + 2k\pi)}\right)^{1/5}$

where $k = 0, \pm 1, \pm 2$.

$Z_1 = \underline{/3\pi/10}$ $\qquad Z_2 = \underline{/3\pi/10 + 2\pi/5}$ $\qquad Z_3 = \underline{/3\pi/10 - 2\pi/5}$

$Z_4 = \underline{/3\pi/10 + 4\pi/5}$ $\qquad Z_5 = \underline{/3\pi/10 - 4\pi/5}$

$Z_1 = \underline{/54^{\circ}}$ $\qquad Z_2 = \underline{/54^{\circ} + 72^{\circ}} = \underline{/126^{\circ}}$

$Z_3 = \underline{/54^{\circ} - 72^{\circ}} = \underline{/-18^{\circ}}$ $\qquad Z_4 = \underline{/54^{\circ} + 144^{\circ}}$, $Z_4 = \underline{/198^{\circ}}$

$Z_5 = \underline{/54^{\circ} - 144^{\circ}} = \underline{/-90^{\circ}}$

Therefore

$Z_1 = 1\underline{/54^{\circ}} = \cos 54^{\circ} + i\sin 54^{\circ}$

$Z_2 = 1\underline{/126^{\circ}} = \cos 126^{\circ} + i\sin 126^{\circ}$

$Z_3 = 1\underline{/342^{\circ}} = \cos 342^{\circ} + i\sin 342^{\circ}$

$Z_4 = 1\underline{/198^{\circ}} = \cos 198^{\circ} + i\sin 198^{\circ}$

$Z_5 = 1\underline{/270^{\circ}} = \cos 270^{\circ} + i\sin 270^{\circ}$
$= -i$

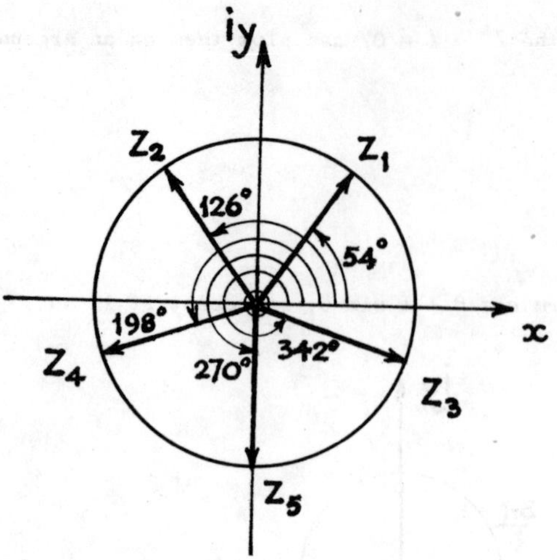

Fig.15 $Z^5 + i = 0$. The moduli of Z_1, Z_2, Z_3, Z_4, Z_5 are equal. The arguments of these complex numbers are 54°, 126°, 342°, 198°, 270° respectively in the range $0 \leqslant \theta \leqslant 360^o$.

The magnitude of all these vectors is unity, and therefore a circle with radius equal to unity is drawn and the angles are measured from the reference in an anticlockwise direction. If, however, the angles are given as $-\pi \leqslant \arg Z \leqslant \pi$, then the complex numbers are as follows:

$Z_1 = \cos 54^o + i \sin 54^o$

$Z_2 = \cos 126^o + i \sin 126^o$

$Z_3 = \cos 18^o - i \sin 18^o$

$Z_4 = \cos 162^o - i \sin 162^o$

$Z_5 = \cos 90^o - i \sin 90^o = -i$

Fig.15 and Fig.16 show respectively, the positive angles and the principal values respectively.

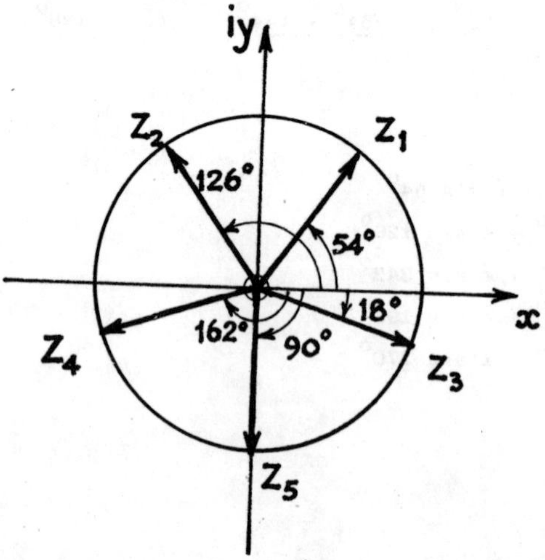

Fig.16 Principal values of the complex numbers, in the range $-180^o \leqslant \theta \leqslant 180^o$, $Z^5 + i = 0$

WORKED EXAMPLE 29

Determine the principal value of $(1 + i)^{3/5}$ and its other values.

....

SOLUTION 29

Let $Z = 1 + i$

$|Z| = \sqrt{2}$

$\arg Z = \tan^{-1} 1 = \pi/4$

$1 + i = \sqrt{2}\ (\cos \pi/4 + i \sin \pi/4)$

The principal value of $(1 + i)^{3/5}$ is

$$\left(\sqrt{2}\right)^{3/5}\left[\cos \frac{3\pi}{20} + i \sin \frac{3\pi}{20}\right]$$

where $k = 0$

The other values can be expressed as

$$\sqrt[10]{8}\left[\cos\left(\frac{3\pi}{20} + \frac{2k\pi}{5}\right) + i \sin\left(\frac{3\pi}{20} + \frac{2k\pi}{5}\right)\right] \text{ where } k = 0, 1, 2, 3, 4 \text{ or } k = 0, \pm 1, \pm 2$$

There are five values in all.

WORKED EXAMPLE 30

Solve $(Z - 1)^n = Z^n$

SOLUTION 30

Taking the nth root on each side

$(Z - 1)^n = Z^n \times 1 \quad \therefore \quad (Z - 1) = Z(1)^{1/n}$

$$Z - 1 = Z\left(\cos \frac{2k\pi}{n} + i \sin \frac{2k\pi}{n}\right) \text{ where } k = 0, 1, 2....(n - 1)$$

since the nth roots of unity are $\cos \frac{2k\pi}{n} + i \sin \frac{2k\pi}{n}$

$$Z\ 1 - \cos \frac{2k}{n} - i \sin \frac{2k}{n} = 1 \qquad \text{where } 2 \sin^2 \frac{k\pi}{n} = \left(1 \cos \frac{2k\pi}{n}\right)$$

$$Z\left(2 \sin \frac{k\pi}{n} - 2i \sin \frac{k\pi}{n} \cos \frac{k\pi}{n}\right) = 1 \qquad \sin \frac{2k\pi}{n} = 2 \sin \frac{k\pi}{n} \cos \frac{k\pi}{2}$$

$$2Z \sin \frac{k\pi}{n}\left(\sin \frac{k\pi}{n} - i \cos \frac{k\pi}{n}\right) = 1$$

Multiplying each side by the $\sin \frac{k\pi}{n} + i \cos \frac{k\pi}{n}$

$$2Z \sin \frac{k\pi}{n} = \sin \frac{k\pi}{n} + i \cos \frac{k\pi}{n}$$

since $\left(\sin \frac{k\pi}{n} - i \cos \frac{k\pi}{n}\right) \times \left(\sin \frac{k\pi}{n} + i \cos \frac{k\pi}{n}\right) = 1$

then $\boxed{Z = \frac{1}{2}\left(1 + i \cot \frac{k\pi}{n}\right)}$ where $k = 0, 1, 2.......(\ \ - 1)$

$P(Z) = (Z - 2 - 3i)\ (Z^3 + (-1 + 3i)Z^2 + (-4 + 3i)Z + (4 - 6i) = 0$

then $Z^3 + (-1 + 3i)Z^2 + (-4 + 3i)Z + (4 - 6i) = 0$

This is rather difficult; since the linear factors have real coefficients we try simple real numbers.

Let $Z = 1$

$P(1)\ 1 - 3 + 7 + 21 - 26 = 0$

Therefore $Z - 1$ is a factor.

Let $Z = -2$

$$\begin{aligned} P(-2) &= (-2)^4 - 3(-2)^3 + 7(-2)^2 + 21(-2) - 26 \\ &= 16 + 24 + 28 - 42 - 26 \\ &= 0 \end{aligned}$$

Therefore $Z + 2$ is another factor.

$(Z - 1)\ (Z + 2) = Z^2 - Z + 2Z - 2 = Z^2 + Z - 2$

$$\begin{array}{r|l} & Z^2 - 4Z + 13 \\ \hline Z^2 + Z - 2 & Z^4 - 3Z^3 + 7Z^2 + 21Z - 26 \\ & Z^4 + Z^3 - 2Z^2 \\ \hline & -4Z^3 + 9Z^2 + 21Z - 26 \\ & -4Z^3 - 4Z^2 + 8Z \\ \hline & 13Z^2 + 13Z - 26 \\ & 13Z^2 + 13Z - 26 \\ \hline & 0 \end{array}$$

Dividing $P(Z)$ by $Z^2 + Z - 2$, gives $Z^2 - 4Z + 13)$

therefore $P(Z) = (Z - 1)(Z + 2)(Z^2 - 4Z + 13) = 0$

$$Z^2 - 4Z + 13 = 0$$

$$Z = (4 \pm \sqrt{16 - 52})/2 = \frac{4 \pm 6i}{2} = 2 \pm 3i$$

Since $2 + 3i$ is a root, then the conjugate of $2 + 3i$, $2 - 3i$ is another root. but since $P(Z)$ has a root $2 + 3i$. it has also its conjugate since the coefficients of $P(Z)$ are real.

$(Z - 2 - 3i)$ and $(Z - 2 + 3i)$ are both factors of $P(Z)$
$(Z - 2 - 3i)\ (Z - 2 + 3i) = (Z - 2)^2 + 9 = Z^2 - 4Z + 13$ which can be found much quicker.

Dividing $P(Z)$ by $Z^2 - 4Z + 13$, it gives $Z^2 + Z - 2$ which factorises easily to $(Z - 1)$ and $(Z + 2)$.

Therefore, $P(Z) = (Z - 1)\ (Z + 2)\ (Z^2 - 4Z + 13)$.

WORKED EXAMPLE 31

Given that $2 + 3i$ is a root of the polynomial equation $P(Z) = 0$, where $P(Z) = Z^4 - 3Z^3 + 7Z^2 + 21Z - 26$, fsctorize $P(Z)$ into linear and quadratic factors with real coefficients.
Find the other 3 roots of the equation $P(Z) = 0$.

.....

SOLUTION 31

$Z = 2 + 3i$, since this is a root of $P(Z)$, then $P(2 + 3i) = 0$

$$Z^3 + (3i - 1)Z^2 + (3i - 4)Z + (4 - 6i)$$

$$Z - 2 - 3i \quad \overline{\big) \; Z^4 - 3Z^3 + 7Z^2 + 21Z - 26}$$

$$Z^4 - 2Z^3 - 3iZ^3$$

$$-Z^3 + 3iZ^3 + 7Z^2 + 21Z - 26$$

$$-Z^3 + 3iZ^3 - 6iZ^2 + 2Z + 9Z^2 + 3iZ^2$$

$$5Z^2 - 9Z^2 - 3iZ^2 + 6iZ^2 + 21Z - 26$$

$$= -4Z^2 + 3iZ^2 + 21Z - 26$$

$$-4Z^2 + 3iZ^2 - 6iZ + 8Z + 9Z + 12iZ$$

$$4Z + 6iZ - 12iZ - 26$$

$$= 4Z - 6iZ - 26$$

$$4Z - 6iZ - 8 + 12i - 12i - 18$$

$$0.$$

Dividing $P(Z)$ by $Z - 2 - 3i$, it gives a zero remainder as is seen above.
This of course is not required entirely.

WORKED EXAMPLE 32

Solve the equation:

$$f(Z) = Z^3 + (2 - i)Z^2 + (5 - 2i) - 5i = 0 \quad .... (1)$$

SOLUTION 32

Let $Z = i$

$$f(i) = i^3 + (2 - i)i^2 + (5 - 2i)i - 5i$$
$$= i - 2 + i + 5i + 2 - 5i$$
$$f(i) = 0$$

One root of the equation (1) is therefore $Z = i$, or $Z - i$ is a factor.

To find the other two roots

Let $z^3 + (2 - i)z^2 + (5 - 2i)z - 5i = (z^2 - i)(z^2 + az + b) = 0$

$$= z^3 + az^2 + bz - iz^2 - iaz - ib$$
$$= z^3 + (a - i)z^2 + (b - ia)z - ib$$

Equating coefficients

$$a - i = 2 - i$$

$$\boxed{a = 2}$$

$$5 - 2i = b - ia$$

$$5 - 2i = b - i2$$

$$\boxed{b = 5}$$

This checks that $-5i = -ib$ from $b = 5$

$$\boxed{Z = i} \qquad z^2 + 2z + 5 = 0$$

$$Z = \frac{2 \pm \sqrt{4 - 20}}{2} = -1 \pm 2i$$

The three roots of the polynomial are

$$\boxed{Z_1 = -1 + 2i}$$

$$\boxed{Z_2 = -1 - 2i}$$

$$\boxed{Z_3 = i}$$

WORKED EXAMPLE 33

Find the roots of the quadratic equation $z^2 - 4z + 8 = 0$, z_1 and z_2 and find their sum and product.

Find $Re\,(z_1{}^6)$ and $Im(z_2{}^8)$.

SOLUTION 33

$z^2 - 4z + 8 = 0$

Solving this quadratic equation $Z = \dfrac{4 \pm \sqrt{16 - 32}}{2} = 2 \pm i2$

The roots are:

$$z_1 = 2 + i2$$
$$z_2 = 2 - i2$$

Then the sum and product of the roots are

$z_1 + z_2 = 4$ and $z_1 z_2 = 8$ respectively.

The moduli of z_1 and z_2 can be found

$$|z_1| = \sqrt{2^2 + 2^2} = 2\sqrt{2}$$

$$|z_2| = \sqrt{2^2 + (-2)^2} = 2\sqrt{2}$$

The arguments of z_1 and z_2 can also be found.

$$\arg z_1 = \tan^{-1} 2/2 = \tan^{-1}1 = \pi/4$$

$$\arg z_2 = -\tan^{-1}2/2 = -\tan^{-1}1 = -\pi/4$$

$$z_1 = 2\sqrt{2}\underline{/\pi/4} = 2\sqrt{2}\,e^{i\pi/4} = 2^{3/2}\,e^{i\pi/4}$$

$$z_1{}^6 = 2^9\,e^{i3\pi/2} = 2^9\,\underline{/3\pi/2} = 2^9\,(\cos 3\pi/2 + i \sin 3\pi/2) = -2^9 i$$

$$Re\,(z_1{}^6) = 0$$

$$Im(z_2{}^8) = Im\,(2\sqrt{2}\underline{/-\pi/4})^8 = Im(2^{12}\,\underline{/-2\pi}) = Im(2^{12}\,\underline{/0^\circ}) = 0$$

$$Re(z_1{}^8) = 2^{12}\,e^{i2\pi} = 2^{12}\,(\cos 2\pi + i \sin 2\pi) = 2^{12}$$

$$Im(z_1)^6 = Im 2^9\,e^{i3\pi/2} = Im\left[2^9 \cos 3\pi/2 + i2^9 \sin 3\pi/2\right]$$

$$Im(z_1)^6 = -2^9$$

Therefore the real part of $z_1{}^6$, namely $Re(z_1{}^6)$ is zero, and the imaginary part of $z_2{}^8$, namely $Im(z_2{}^8)$ is zero

WORKED EXAMPLE 34

The roots of a polynomial are $z = -3$, $z = 3 - i$, and $z = 3 + i$.

Determine the polynomial equation.

.....

SOLUTION 34

The factors of the polynomial equation are $(z + 3)$, $(z - 3 + i)$ and $(z - 3 - i)$, therefore the polynomial equation will be

$$(z + 3)\ (z - 3 + i)\ (z - 3 - i) = 0 \quad \ldots\ldots\ldots\ldots \quad (1)$$

from which we deduce that $z + 3 = 0$, $z - 3 + i = 0$ and $z - 3 - i = 0$ or $z = -3$, $z = 3 - i$, and $z = 3 + i$.

Multiplying out equation (1)

$$(z + 3)\left[(z - 3) + i\right] . \left[(z - 3) - i\right] = (z + 3)\left[(z - 3)^2 + 1\right] = 0$$

$$= (z + 3)(z^2 - 6z + 9 + 1 = 0$$

$$= z^3 - 6z^2 + 10z + 3z^2 - 18z + 30 = 0$$

$$\text{or } z^3 - 3z^2 - 8z + 30 = 0$$

It is observed that the polynomial equation has real coefficients since the roots appear in conjugate pairs.

WORKED EXAMPLE 35

Now try the following question:

The roots of a cubic equation in Z are as follows:

$Z = 1$, $Z = 3 - i4$ and $Z = 3 + i4$.

Determine the equation.

SOLUTION 35

$(Z - 1).\ (Z - 3 + i4)\ .\ (Z - 3 - i4) = 0$, the product of the factors is equal to zero.

$$(Z - 1)\left[(Z - 3) + i4\right]\left[(Z - 3) - i4\right] = 0$$

$$(Z - 1)\left[(Z - 3)^2 + 16\right] = (Z - 1)(Z^2 - 6Z + 25) = Z^3 - 6Z^2 + 25Z - Z^2 + 6Z - 25$$

$$= Z^3 - 7Z^2 + 31Z - 25 = 0$$

It is quite easy to formulate the complex polynomial with real coefficients.

WORKED EXAMPLE 36

Now try to think how you are going to solve the polynomial $Z^3 - 3Z^2 - 8Z + 30 = 0$ showing that one root is $Z = 3 - i$, in other words, given one complex root, find the other two roots.

SOLUTION 36

The problem is again easy, but this time the technique is different.

Knowing that $Z = 3 - i$, then another root is $Z = 3 + i$, the conjugate of $Z = 3 - i$, since the polynomial has real coefficients, we know that the roots appear in conjugate pairs.

Therefore, the two roots are $Z = 3 - i$ and $Z = 3 + i$ or their factors are $(Z - 3 + i)$ and $(Z + 3 - i)$.
Multiplying $(Z - 3 + i)\ (Z - 3 - i) = \left[(Z - 3) + i\right]\left[(Z - 3) - i\right]$
$= (Z - 3)^2 + 1 = Z^2 - 6Z + 10.$

Hence to find the third root we divided the given polynomial $Z^3 - 3Z^2 - 8Z + 30 = 0$ by $Z^2 - 6Z + 10$.

$$
\begin{array}{r|l}
 & Z + 3 \\
\hline
Z^2 - 6Z + 10 & Z^3 - 3Z^2 - 8Z + 30 \\
 & Z^3 - 6Z^2 + 10Z \\
\hline
 & \quad 3Z^2 - 18Z + 30 \\
 & \quad 3Z^2 - 18Z + 30 \\
\hline
 & \quad - \quad - \quad -
\end{array}
$$

Therefore, the roots of the polynomial equation are $Z = -3$, $Z = 3 - i$, and $Z = 3 + i$ and the factors

$$(Z + 3)(Z - 3 + i)(Z - 3 - i) = Z^3 - 3Z^2 - 8Z + 30 = 0$$

WORKED EXAMPLE 37

Now try and solve the following problem:

If $Z = 3 + i4$ is a root of the equation $Z^3 - 7Z^2 + 31Z - 25 = 0$

find the other roots.

SOLUTION 37

Since $Z = 3 + i4$ is a root of the polynomial equation $Z^3 - 7Z + 31Z - 25 = 0$ another root is the conjugate of $Z = 3 + i4$, namely $Z = 3 - i4$.

The factors of $Z = 3 + i4$ and $Z = 3 - i4$ are $(Z - 3 - i4)$ and $(Z - 3 + i4)$ The product of these factors are equal to zero since each is equal to zero, being the root of the polynomial.

$$\left(Z - 3 - i4\right)\left(Z - 3 + i4\right) = 0$$

$$(Z - 3)^2 - (i4)^2 = 0 \text{ or } Z^2 - 6Z + 9 + 16 = Z^2 - 6Z + 25 = 0$$

To find the third root, we divide the polynomial by the quadratic factor.

$$\begin{array}{r|l} & Z - 1 \\ \hline Z^2 - 6Z + 25 & Z^3 - 7Z^2 + 31Z - 25 \\ & Z^3 - 6Z^2 + 25Z \\ \hline & -Z^2 + 6Z - 25 \\ & -Z^2 + 6Z - 25 \\ \hline & = \quad = \quad = \end{array}$$

Therefore the three roots of the polynomial are $Z = 1, Z = 3 + i4$ and $Z = 3 - i4$, and the polynomial can be expressed

$$Z^3 - 7Z^2 + 31Z - 25 = (Z - 1)(Z - 3 - i4)(Z - 3 + i4) = 0$$

$$= (Z - 1)(Z^2 - 6Z + 25) = 0$$

Quadratic equations can easily be formed with real coefficients, knowing the complex number and of course the conjugate complex number can be written down.

(a) If $Z = i$, its conjugate is $Z = -i$

$(Z - i)(Z + i) = Z^2 - i^2 = Z^2 + 1 = 0$

The required quadratic equation is $\boxed{Z^2 + 1 = 0}$

(b) If $Z = -1 - 2i$, its conjugate is $Z = -1 + 2i$ and the quadratic equation can be found by writing down the factors and multiplying them out.

$$(Z + 1 + 2i)(Z + 1 - 2i) = \left[(Z + 1) + 2i\right]\left[Z + 1) - 2i\right]$$

$$= (Z + 1)^2 - 4i^2 = Z^2 + 2Z + 1 + 4$$

$$\therefore \boxed{Z^2 + 2Z + 5 = 0}$$

(c) If $Z = -5 + 7i$, determine the quadratic equation with real coefficients. The conjugate complex number is $Z = -5 - 7i$

$$(Z + 5 - 7i)(Z + 5 + 7i) = 0$$

$$(Z + 5)^2 - 49i^2 = Z^2 + 10Z + Z + 49 = Z^2 + 10Z + 51 = 0$$

$$\therefore Z^2 + 10Z + 51 = 0$$

One root of the quadratic equations:

(i) $Z^2 + 1 = 0$ ($Z = -i$)

(ii) $Z^2 + 2Z + 5 = 0$ ($Z = -1 + 2i$)

(iii) $Z^2 + 10Z + 51 = 0$ ($Z = -5 - 7i$)

As shown adjacent to each equation, find the other root.

The answers are quite easy now, for (i) $Z = i$
(ii) $Z = -1 - 2i$
(iii) $Z = -5 + 7i$

WORKED EXAMPLE 38

Find the five roots of $Z^5 - 32 = 0$, and write down linear and quadratic factors of this equation with real coefficients.

SOLUTION 38

$Z^5 - 32 = 0 \qquad Z = 2\,(1)^{1/5} = 2\,(\cos 2k\pi + i \sin 2k\pi)^{1/5}$

$Z^5 = 2^5 \qquad$ where $k = 0, \pm 1, \pm 2$.

$Z = 2, \quad Z = 2\underline{/\ \pm 2\ \pi/5} \quad$ and $\quad Z = 2\underline{/\ \pm 4\pi/5}$

$(Z - 2).(Z - 2\cos 2\pi/5 - 2i \sin 2\pi/5).(Z - 2\cos 2\pi/5 + 2i \sin 2\pi/5).$

$(Z - 2\cos 4\pi/5 - 2i \sin 4\pi/5).(Z - 2\cos 4\pi/5 + 2i \sin 4\pi/5) = 0$

$(Z - 2)\ (Z - 2\cos 2\pi/^5)^2 - 4i^2 \sin^2 2\pi/5)\ (Z - 2\cos 4\pi/5)^2 - 4i^2 \sin^2 4\pi/5) = 0$

$(Z - 2)\ (Z^2 - 4Z\cos 2\pi/5 + 4\cos^2 2\pi/5 + 4\sin^2 2\pi/5).$

$\qquad (Z - 4Z\cos 4\pi/5 + 4\cos^2 4\pi/5 + 4\sin^2 4\pi/5) = 0$

$(Z - 2)\ (Z^2 - 4Z\cos 2\pi/5 + 4)\ (Z^2 - 4Z\cos 4\pi/5 + 4) = 0$

Again we observed that the roots appear in conjugate pairs since the coefficients of $Z^5 - 32 = 0$ are real.

WORKED EXAMPLE 39

If $Z = 5 + i12$, $Z = -3 - i4$, and $Z = -2$ are the other roots with real coefficients of a polynomial equation of degree five, *determine the polynomial.*

...

SOLUTION 39

Since $Z = 5 + 12i$, then the conjugate root is $Z = 5 - 12i$, and since $Z = -3 - i4$, then the conjugate root is $Z = -3 + i4$. The polynomial is determined as follows:

Since the roots are now given as:

$Z = -2 \qquad\qquad Z = -3 - i4$

$Z = 5 + 12i \qquad Z = -3 + i4$

$Z = 5 - 12i$

These form the factors:

$(Z + 2)\ (Z - 5 + 12i)\ (Z - 5 - 12i)\ (Z + 3 - i4)\ (Z + 3 + i4) = 0$

$(Z + 2)\ \left[(Z - 5)^2 - 12^2\ i^2\right]\ \left[(Z + 3)^2 - i^2\ 4^2\right] = 0$

$(Z + 2)\ \left[Z^2 - 10Z + 25 + 144\right]\left[Z^2 + 6Z + 9 + 16\right] = 0$

$(Z + 2)(Z^4 - 10Z^3 + 169Z^2 + 6Z^3 - 60Z^2 + 1014Z + 25Z^2 - 250Z + 4225) = 0$

$(Z + 2)\ (Z^4 - 4Z^3 + 134Z^2 + 764Z + 4225) = 0$

$Z^5 - 4Z^4 + 134Z^3 + 764Z^2 + 4225Z + 2Z^4 - 8Z^3 + 268Z^2 + 1528Z + 8450 = 0$

$Z^5 - 2Z^4 + 126Z^3 + 1032Z^2 + 5753Z + 8450 = 0$

WORKED EXAMPLE 40

Find the four roots of the equation

$$Z^4 - 8Z^3 + 34Z^2 - 72Z + 65 = 0$$

given that one root is $Z = 2 - i$.

SOLUTION 40

The sum of the roots $\alpha + \beta + \gamma + \delta = 8$, and their product $\alpha\beta\gamma\delta$ is 65.

Since $\alpha = 2 - i$, then $\beta = 2 + i$ since the roots appear in conjugate pairs because the polynomial given has real coefficients.

The sum of these roots are $\alpha + \beta = 4$, and their product $\alpha\beta = (2 - i)(2 + i) = 4 + 1 = 5$.
Therefore $\gamma + \delta = 8 - 4 = 4$ and $\gamma\delta = \frac{65}{5} = 13$.
Thus $\gamma^2 - 4\gamma + 13 = 0$, solving this quadratic gives $\gamma = 2 + 3i$ and $\delta = 2 - 3i$.

WORKED EXAMPLE 41

Given that $|Z + 2| = 1$

sketch the logus of a point $P(x, y)$ which represents Z on an Argand diagram.

SOLUTION 41

$|Z + 2| = 1$

Let $Z = x \quad iy$

$|x + iy + 2| = 1 \quad \sqrt{(x + 2)^2 + y^2} = 1$

Squaring up both sides

$(x + 2)^2 + y^2 = 1$ or $x^2 + 4x + y^2 + 4 - 1 = 0$

The locus is a circle with centre $c(-2,0)$ and radius $r = 1$.

Fig. 17 shows this locus.

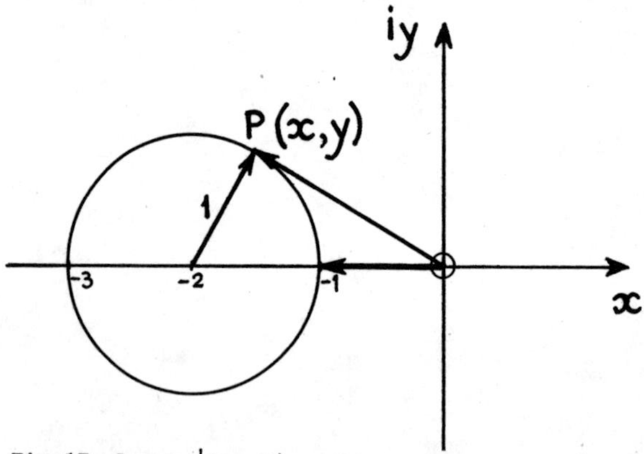

Fig.17 Locus $|z + 2| = 1$ is a circle, $c(-2,0)$, $r = 1$.

The cartesian equation of the circle is $x^2 + y^2 + 4x + 3 = 0$

and the complex number form is $|Z + 2| = 1$

WORKED EXAMPLE 42

Describe the locus $|Z + 1 - 2i| = 5$

SOLUTION 42

Let $Z = x + iy$

$|x + iy + 1 - 2i| = 5$

$|x + 1 + i(y - 2)| = 5 \quad \sqrt{(x + 1)^2 + (y - 2)^2} = 5$

and squaring up $(x + 1)^2 + (y - 2)^2 = 5^2$

The locus is a circle with a centre $c(-1, 2)$ and radius $r = 5$

Fig. 18 shows this locus.

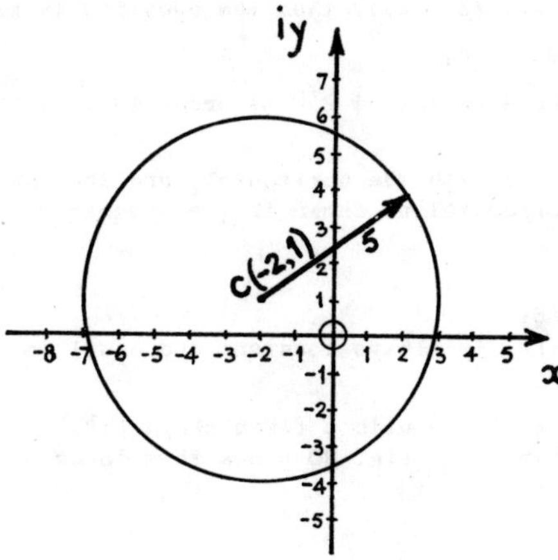

Fig.18 Locus $|z + 1 - 2i| = 5$ is a circle, $C(-2,1)$ and $r = 5$.

THE LOCUS $\arg\left(\frac{Z - Z_1}{Z - Z_2}\right) = \theta_1$

It is required to find the locus of $\arg\left(\frac{Z - Z_1}{Z - Z_2}\right) = \theta_1$
where $Z = x + iy$ (variable), $Z_1 = x_1 + iy_1$ (fixed) and $Z_2 = x_2 + iy_2$ (fixed) and θ_1 is an acute angle ($0 < \theta_1 < \pi$).

Let the complex numbers Z, Z_1 and Z_2 be represented by the points P, P_1 and P_2 respectively.

$\arg\left(\frac{Z - Z_1}{Z - Z_2}\right) = \arg(Z - Z_1) - \arg(Z - Z_2) = \theta_1$

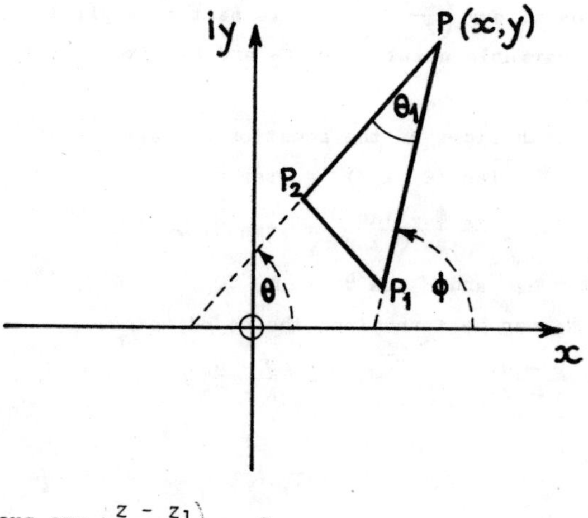

Fig.19 The locus $\arg\left(\frac{z - z_1}{z - z_2}\right) = \theta_1$

Fixed points $P_2(x_2, y_2)$ and $P_1(x_1, y_1)$.
Variable point $P(x, y)$

Let $\phi = \arg(Z - Z_1)$ and $\theta = \arg(Z - Z_2)$, then the equation is simplified to

$$\phi - \theta = \theta_1$$

From this equation, it can be seen that $\phi > \theta$ in order that θ_1 is a positive angle.

The vector P_1P makes an angle ϕ with the horizontal, and the vector P_2P makes an angle θ with the horizontal as shown in the diagram of Fig. 19.

$$\therefore \quad \phi - \theta = \theta_1$$

The angle P_1PP_2 is equal to θ_1
The point P moves so that $P_1\hat{P}P_2$ is always constant and equal to θ_1, the angle given in the problem.

The locus of P is an arc of a circle with a fixed chord P_1P_2 subtending an angle θ_1 at the circumference. Fig. 20 shows this locus.

$$\phi - \theta = \theta_1$$

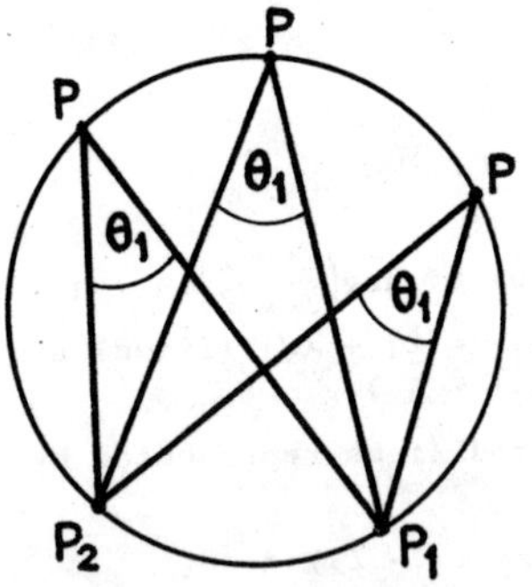

Fig.20 The locus of $\arg\left(\frac{z - z_1}{z - z_2}\right) = \theta$ is part of a circle.
$P(x, y)$ variable point, with P_2 and P_1 fixed points.

Taking tangents on both sides of the equation we have

$$\tan(\phi - \theta) = \tan\theta_1$$

$$\frac{\tan\phi - \tan\theta}{1 + \tan\phi\tan\theta} = \tan\theta_1$$

$$\tan\phi = m_2 \quad \text{and} \quad \tan\theta = m_1$$

The gradients m_1, m_2 can be found from the coordinates.

$$m_2 = \frac{y - y_1}{x - x_1} \qquad m_1 = \frac{y - y_2}{x - x_2}.$$

WORKED EXAMPLE 43

Determine the locus of Z if $\arg\left(\frac{Z - Z_1}{Z - Z_2}\right) = \pi/4$

SOLUTION 43

$\arg\left(\frac{Z - Z_1}{Z - Z_2}\right) = \arg(Z - Z_1) - \arg(Z - Z_2) = \pi/4$
from which we have $\phi - \theta = \pi/4$ if $\phi = \arg(Z - Z_1)$ and $\theta = \arg(Z - Z_2)$.

$$\tan(\phi - \theta) = \tan \pi/4$$

$$\frac{\tan\phi - \tan\theta}{1 + \tan\phi\tan\theta} = 1$$

$$\frac{m_2 - m_1}{1 + m_1 m_2} = 1$$

$$\frac{\frac{y - y_1}{x - x_1} - \frac{y - y_2}{x - x_2}}{1 + \frac{y - y_1}{x - x_1} \cdot \frac{y - y_2}{x - x_2}} = 1 \quad \therefore (y - y_1)(x - x_2) - (y - y_2)(x - x_1)$$
$$= (x - x_1)(x - x_2) + (y - y_1)(y - y_2)$$

If $Z_1 = -1$ and $Z_2 = 1$ i.e. $x_1 = -1$, $y_1 = 0$, $x_2 = 1$ and $y_2 = 0$

$(y - y_1)(x - x_2) - (y - y_2)(x - x_1) = (x - x_1)(x - x_2) + (y - y_1)(y - y_2)$
can be written

$$y(x - 1) - y(x + 1) = (x + 1)(x - 1) + y^2$$

$$y^2 + x^2 - 1 - yx + y + yx + y = 0$$

$$y^2 + x^2 + 2y - 1 = 0 \quad \text{or} \quad (y + 1)^2 + x^2 = 2$$

This is a circle with centre $C(0, -1)$ and radius $r = \sqrt{2}$.

$P_1\hat{P}P_2 = 45^\circ = \pi/4$

The locus is the major arc of the circle shown in the diagram P_1PP_2, Fig. 21.

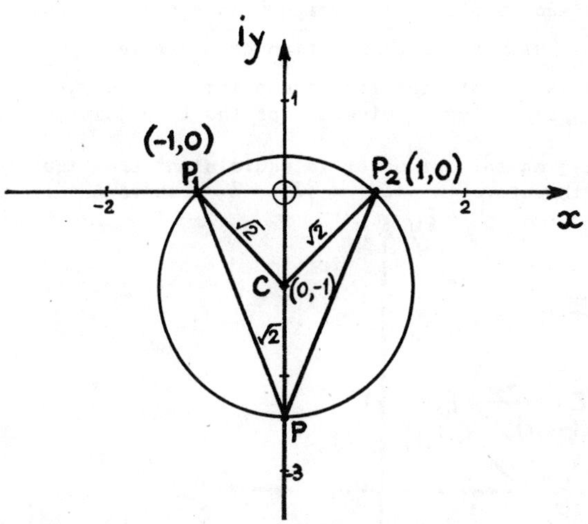

Fig.21 The locus is part of a circle.
The major part of a circle $C(0, -1)$, $r = \sqrt{2}$.

WORKED EXAMPLE 44

Determine the locus of Z given by the equation

$$\arg\left(\frac{Z + 1 - i}{Z + 2}\right) = \pi/4$$

and sketch it carefully on an Argand diagram.

SOLUTION 44

$Z_1 = -1 + i$ and $Z_2 = -2$

$$\arg\left(\frac{Z + 1 - i}{Z + 2}\right) = \pi/4 \quad \text{or} \quad \arg(Z + 1 - i) - \arg(Z + 2) = \pi/4$$

$$\arg(x + 1 + i(y - 1)) - \arg(x + 2 + iy) = \pi/4$$

$$\tan^{-1}\frac{y - 1}{x + 1} - \tan^{-1}\frac{y}{x + 2} = \pi/4$$

Take the tangent on both sides

$$\frac{\tan\tan^{-1}\frac{y - 1}{x + 1} - \tan\tan^{-1}\frac{y}{x + 2}}{1 + \tan\tan^{-1}\frac{y - 1}{x + 1} \cdot \tan\tan^{-1}\frac{y}{x + 2}} = \tan \pi/4$$

$$\frac{\frac{y - 1}{x + 1} - \frac{y}{x + 2}}{1 + \frac{(y - 1)(y)}{(x + 1)(x + 2)}} = 1$$

$$\frac{(y - 1)(x + 2) - y(x + 1)}{(x + 1)(x + 2)} = 1 + \frac{(y - 1)y}{(x + 1)(x + 2)}$$

$$xy - x + 2y - 2 - yx - y = x^2 + 3x + 2 + y^2 - y$$

$$x^2 + y^2 + 4x - 2y + 4 = 0$$

$$(x + 2)^2 - 4 + (y - 1)^2 - 1 + 4 = 0$$

$$(x + 2)^2 + (y - 1)^2 = 1^2$$

This is the locus which is a circle with centre C (-2, 2) and radius $r = 1$
Fig. 22 shows this locus.

Describe the locus of Z given that $\left|\frac{Z - Z_1}{Z - Z_2}\right| = k$ ...... (1)

where Z_1 and Z_2 are fixed complex numbers and k is a positive constant.

Equation (1) represents either a straight line or a circle.

If $k = 1$, the point Z is equidistant from the points Z_1 and Z_2 and therefore lies on the perpendicular bisector of the line joining these points.
Conversely, any point Z on this bisector is equidistant from the points Z_1 and Z_2 and therefore $|Z - Z_1| = |Z - Z_2|$ where $k = 1$.

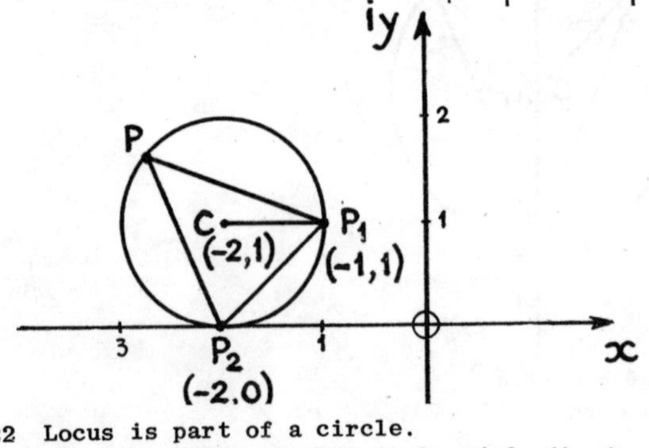

Fig.22 Locus is part of a circle.
The major part of the circle C(-2, 1), $r = 1$.

WORKED EXAMPLE 45

Describe the locus of Z given that $|Z - Z_1| = |Z - Z_2|$
or $|Z - (3 + i4)| = |Z - (1 + i2)|$
where $Z_1 = 3 + i4$ and $Z_2 = 1 + i2$, the fixed complex numbers.

SOLUTION 45

The point Z is equidistant from the points Z_1 and Z_2 which are represented by P_1 and P_2 respectively, so OP_1 and OP_2 are the vectors Z_1 and Z_2.
Fig. 23 shows these points $|Z - Z_1| = |Z - Z_2|$
Substituting $Z = x + iy$ and $Z_1 = 3 + i4$, $Z_2 = 1 + i2$, we have

$$|x + iy - (3 + i4)| = |x + iy - (1 + i2)|$$
$$\sqrt{(x - 3)^2 + (y - 4)^2} = \sqrt{(x - 1)^2 + (y - 2)^2}$$

and squaring up both sides, $x^2 - 6x + 9 + y^2 - 8y + 16$
$= x^2 - 2x + 1 + y^2 - 4y + 4$

therefore, the locus $\boxed{x + y = 5}$ is a straight line. Fig. 23.

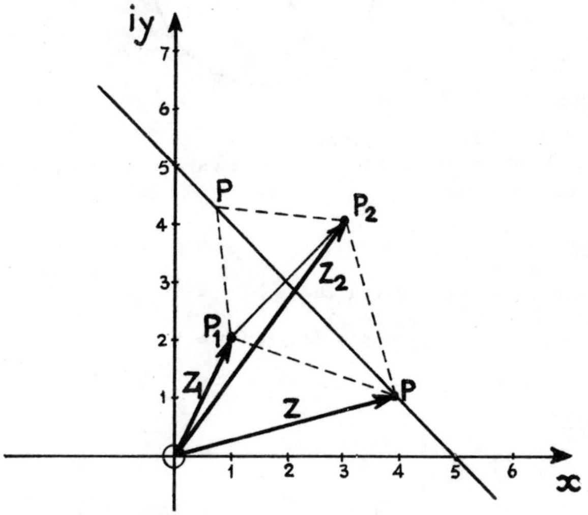

Fig.23 The locus of $|z - (3 + i4)| = |z - (1 + i2)|$.
The locus is a straight line $x + y = 5$.
P_1 and P_2 are fixed points.

Therefore, Z is a variable point lying on the straight line $x + y = 5$ which is the perpendicular bisector of the line joining the fixed points P_1 and P_2.

If $\left|\frac{Z - Z_1}{Z - Z_2}\right| = k$ where k is greater than 1, and Z_1 and Z_2 are fixed complex numbers then the equation represents a circle.

A point which moves so that the ratio of its distances from two fixed points P_1 and P_2, is constant, is the locus of a circle with respect to which Z_1 and Z_2 are inverse points. This describes an Apollonius circle, that is, if P_1, P_2 are two fixed points and P is a moving or variable point such that the ratio PP_1 / PP_2 is constant, the locus of P is a circle.

P_1P_2 is divided internally at A and externally at B in the given ratio PP_1/PP_2 since $\frac{PP_1}{PP_2} = \frac{P_1A}{AP_2} = \frac{P_1B}{P_2B}$

PA and PB are the internal and external bisectors of the angle P_1PP_2. Hence the angle APB is a right angle and P therefore lies on the circle whose diameter is AB. This circle is called " the circle of Apollonius".

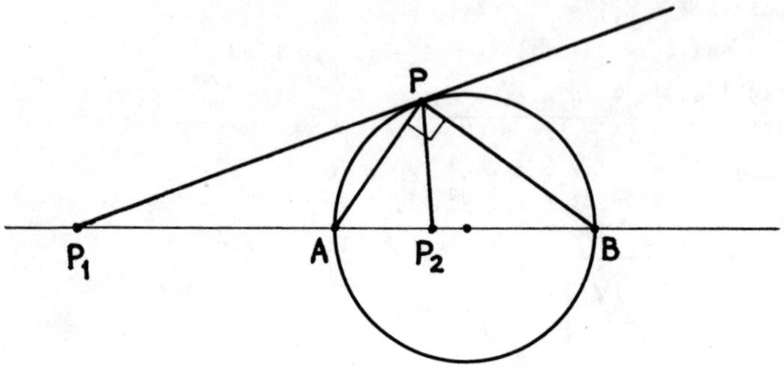

Fig.24 The circle of Apollonius.
P_1 and P_2 are fixed. P is a variable point such as PP_1/PP_2 is constant.

WORKED EXAMPLE 46

Describe the locus of Z given that $\left|\frac{Z - Z_1}{Z - Z_2}\right| = k$

where $Z_1 = 2 + i$ and $Z_2 = 3 + i4$ and $k = 2$

SOLUTION 46

$$\left|\frac{Z - Z_1}{Z - Z_2}\right| = k$$

$$\left|\frac{Z - (2 + i)}{Z - (3 + i4)}\right| = 2 \quad \text{or} \quad \left|\frac{Z - (1 + i)}{Z - (3 + i4)}\right| = 2 \;\ldots\ldots(1)$$

$$|Z - (2 + i)| = 2\,|Z - (3 + i4)|$$

The numerator of equation (1) represents the distance between the point Z and the fixed point $(2 + i)$ or (2,1) and the denominator represents the distance between the point Z and the point $(3 + i4)$ or (3,4).

The distance of Z from (2,1) is therefore twice the distance of Z from the point (3,4), since $k = 2$.

The locus is an Apollonius circle with a centre that lies outside the line joining the points P_1 (2,1) and P_2 (3,4), Fig. 25.

$$|x + iy - 2 - i| = 2\,|x + iy - 3 - i4|$$

$$|(x - 2) + i(y - 1)| = 2\,|(x - 3) + i(y - 4)|$$

$$\sqrt{(x - 2)^2 + (y - 1)^2} = 2\sqrt{(x - 3)^2 + (y - 4)^2}$$

Squaring up and expanding

$$x^2 - 4x + 4 + y^2 - 2y + 1 = 4(x^2 - 6x + 9 + y^2 - 8y + 16)$$

$$3x^2 + 3y^2 - 24x + 4x - 32y + 2y + 100 - 5 = 0$$

$$3x^2 + 3y^2 - 20x - 30y + 95 = 0$$

$$x^2 + y^2 - \frac{20x}{3} - 10y + \frac{95}{3} = 0$$

$$\left(x - \frac{10}{3}\right)^2 - \frac{100}{9} + (y - 5)^2 - 25 + \frac{95}{3} = 0$$

$$\left(x - \frac{10}{3}\right)^2 + (y - 5)^2 = \frac{100}{9} + \frac{75}{3} - \frac{95}{3} = \frac{100}{9} + \frac{225}{9} - \frac{285}{9} = \frac{325 - 285}{9}$$

$$\left(x - \frac{10}{3}\right)^2 + (y - 5)^2 = \frac{40}{9} = \left(\frac{\sqrt{40}}{3}\right)^2$$

The circle of Fig. 25 has a centre $C\left(\frac{10}{3}, 5\right)$ and a radius of $\frac{\sqrt{40}}{3}$

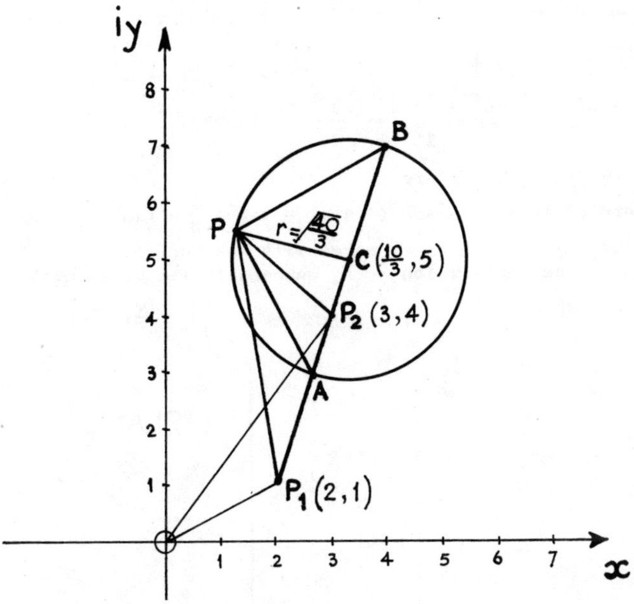

Fig.25 Apollonius circle. $C(^{10}/_3, 5)$, $r = \sqrt{40}/_3$
The locus of z such that $\left|\frac{z - z_1}{z - z_2}\right| = K$
$z_1 = 2 + i$ $z_2 = 3 + i4$, and $K = 2$.

WORKED EXAMPLE 47

The complex numbers Z_1, Z_2 and Z_3 are represented on an Argand diagram by the points P_1, P_2 and P_3 respectively.

If $Z_1 = 1 + i1$, $Z_2 = 5 + i2$ and $Z_3 = 3 + i7$ determine the modulus and argument of $\dfrac{Z_3 - Z_1}{Z_2 - Z_1}$ and represent all these complex numbers on an Argand diagram.

SOLUTION 47

$Z_1 = 1 + i$ $\quad Z_2 = 5 + i2$ $\quad Z_3 = 3 + i7$

OP_1 represents $Z_1 = 1 + i$

OP_2 represents $Z_2 = 5 + i2$

OP_3 represents $Z_3 = 3 + i7$

P_3P_1 represents the vector $Z_3 - Z_1 = 3 + i7 - (1 + i) = 2 + i6$

P_2P_1 represents the vector $Z_2 - Z_1 = 5 + i2 - (1 + i) = 4 + i$

$$Z_2 - Z_1 = 4 + i$$

$$\left|\frac{Z_3 - Z_1}{Z_2 - Z_1}\right| = \left|\frac{2 + i6}{4 + i}\right| = \frac{\sqrt{6^2 + 2^2}}{\sqrt{4^2 + 1^2}} = \frac{\sqrt{40}}{\sqrt{17}} = 1.53$$

$$\arg\left(\frac{Z_3 - Z_1}{Z_2 - Z_1}\right) = \arg(Z_3 - Z_1) - \arg(Z_2 - Z_1)$$

$= \arg(2 + i6) - \arg(4 + i) = \tan^{-1}\frac{6}{2} - \tan^{-1}\frac{1}{4} = 53^\circ 32'$

= the angle which $P_3 P_1$ makes with the horizontal - the angle which $P_2 P_1$ makes with the horizontal

= the angle $P_3 P_1 P_2$

FIG. 26

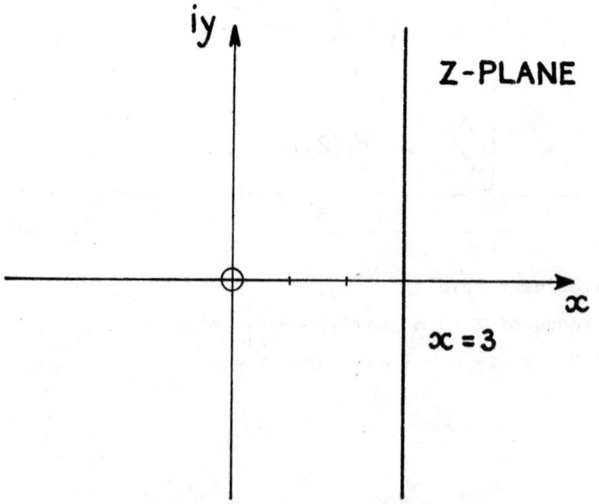

Fig.26 P describes $Re(z) = 3$
Z-plane, the locus of P is $x = 3$.

EXERCISES 17 / 18

1. If P represents the complex number Z, find the loci

   (i) $|Z| = 5$ (iv) $|2Z - 1| = 3$

   (ii) $|Z - 1| = 2$ (v) $|Z - 2 - 3i| = 4$

   (iii) $|Z + 2| = 3$ (vi) $\arg Z = 0$

   (consider $Z = x + iy$).

2. What are the least and greatest values of the following:

   (i) $|Z - 3|$ if $|Z| \leq 1$

   (ii) $|Z + 2|$ if $|Z| \leq 1$

   (iii) $|Z|$ if $|Z - 5| \leq 2$

   (iv) $|Z + 1|$ if $|Z - 4 \leq 3$

   (v) $|Z - 4|$ if $|Z + 3i| \leq 1$.

3. Represent in the Argand diagram

   (i) The cube roots of i
   (ii) The fourth roots of $-i$
   (iii) The fifth roots of -32
   (iv) The cube roots of 8.

4. Solve (i) $(Z + i)^6 + (Z - i)^6 = 0$
   (ii) $(Z + 1)^6 + (Z - 1)^6 = 0$

5. Solve $(Z + 1)^n = (1 - Z)^n$.

6. Use the modulus notation due to Weierstrass to express that the point P which represents the complex number Z lies

   (i) inside the circle with centre (8, 9) and radius 7
   (ii) on the circle with centre (a,b) and radius C
   (iii) outside the circle with centre (-1, 0), radius 1.

   (Ans. (i) $|Z - 8 - 9i| = 4$
   (ii) $|Z - a - bi| = C$
   (iii) $|Z + 1| > 1$

7. Sketch the locus in the Argand diagram of the point representing Z, where $\left|\frac{Z - 1}{Z + 1}\right| = \frac{1}{2}$.

8. If P represents the complex number Z on an Argand diagram, find the cartesian equation of the locus of P

   when $\left|\frac{Z + 1}{Z + 2i}\right| = 5$

9. Write down the fifth roots of -1 and show that $\cos \pi/5 + \cos 3\pi/5 = \frac{1}{2}$

10. If $Z^9 - 1 = 0$, show that $\cos 2\pi/9 + \cos 4\pi/9 + \cos 6\pi/9 + \cos 8\pi/9 = -\frac{1}{2}$

11. If $Z^7 + 1 = 0$, show $\cos \pi/7 + \cos 3\pi/7 + \cos 5\pi/7 = \frac{1}{2}$.

12. Find the six roots of $Z^6 - 2Z^3 + 4 = 0$ and the product of three quadratic factors with real coefficients.

13. Find the roots of $Z^6 - 1 = 0$ and hence the roots of $Z^5 + Z^4 + Z^3 + Z + 1 = 0$.

14. Factorise $Z^4 + 1$.

15. Show that $(Z^n - e^{i\theta})(Z^n - e^{-i\theta}) = Z^{2n} - Z^n \cos\theta + 1$.
Hence, find the roots of the equation

$$Z^8 + Z^4\sqrt{3} + 1 = 0$$

and illustrate these roots in an Argand diagram.

19. COMPLEX FORM OF A CIRCLE AND STRAIGHT LINE

Expresses the circle in complex form

$$|z| = a$$

This is the equation of a circle in complex number, if $z = x + iy$ where x and y are real.

$$|x + iy| = a$$

$$\sqrt{x^2 + y^2} = a$$

Squaring both sides

$$x^2 + y^2 = a^2$$

This is the cartesian form of a circle.

WORKED EXAMPLE 48

A circle has a centre with coordinate (2, -3) and radius 5. Find the complex form of the circle.

SOLUTION 48

$(x - 2)^2 + (x + 3)^2 = 5^2$ is the cartesian form of the circle.

Square rooting $\sqrt{(x - 2)^2 + (y + 3)^2} = 5$ both sides and in complex form

$$|(x - 2) + i(y + 3)| = 5$$

The complex form of the circle is

$$|z - 2 + 3i| = 5$$

This can be written down by observing that the x coordinate of the centre is written down with the opposite sign and the y coordinate of the centre is written down with the opposite sign but with i.

A circle with centre C (-1, -2) and $r = 3$ has a complex form equation.

$$|z + 1 + 2i| = 3$$

The Straight Line in Complex Form

$$|z - z_1| = |z - z_2|$$

$$|x + iy - x_1 - i\,y_1| = |x + iy - x_2 - iy_2|$$

$$|(x - x_1) + i(y - y_1)| = |(x - x_2 + i(y - - y_2)|$$

$$\sqrt{(x - x_1)^2 + (y - y_1)^2}^2 = \sqrt{(x - x_2)^2 + (y - y_2)^2}^2$$

$$x^2 - 2x\,x_1 + x_1^2 + y^2 - 2y\,y_1 + y_1^2$$
$$= x^2 - 2x\,x_2 + x_2^2 + y^2 - 2y\,y_2 + y_2^2, \; - 2x\,x_1 + x_1^2 - 2y\,y_1 + y_1^2$$
$$= -2x\,x_2 + x_2^2 - 2y\,y_2 + y_2^2$$

$$x(2x_1 - 2x_2) + y(2y_1 - 2y_2) + x_2^2 + y_2^2 - x_1^2 - y_1^2 = 0$$

The equation of a straight line.

If $x_1 = 2$, $x_2 = -3$, $y_1 = -1$ and $y_2 = 1$, then

$$x(2 \times 2 - 2 \times (-3) + y(2 \times (-1) - 2 \times 1) + (-3)^2 + 1^2 - 2^2 - (-1)^2 = 0$$

$$10x - 4y + 9 + 1 - 4 - 1 = 0$$

$$10x - 4y + 5$$ is the straight line.

20. TRANSFORMATIONS FROM A Z-PLANE TO A W-PLANE EMPLOYING COMPLEX NUMBERS.

If $z = x + yi$ is represented by the point $P(x,y)$ in the z plane and $w = u + vi$ is represented by the point $Q(u,v)$ in the w plane, then a relationship between z and w defines the mapping of the point P to the point Q.
These transformations are best illustrated by several examples.

TRANSFORMATIONS

WORKED EXAMPLE 49

The points P and Q lie in the z-plane and the w-plane respectively. Given that $z = \frac{1}{w}$, find an equation of the path of Q when P describes the line $Re(x) = 3$.
Represent the paths of P and Q on the z and w planes.

SOLUTION 49

$Re(z) = 3$ describes a straight line in the z-plane
$Re(x + iy) = 3$
therefore $x = 3$ and $y = 0$, but $x + iy = \frac{1}{w}$ where $w = u + iv$

$$\text{then } w = \frac{1}{x + iy} \times \frac{x - iy}{x - iy} = \frac{x}{x^2 + y^2} - i\frac{y}{x^2 + y^2} = u + iv$$

It is required to find an expression connecting u and v, thus eliminating x and y.
Equate real and imaginary terms, squaring up both expressions.

$$u = \frac{x}{x^2 + y^2} \qquad u^2 = \frac{x^2}{(x^2 + y^2)^2} \quad \dots\dots\dots\dots (1)$$

$$v = -\frac{y}{x^2 + y^2} \qquad v^2 = \frac{y^2}{(x^2 + y^2)^2} \quad \dots\dots\dots (2)$$

Adding equations (1) and (2), we have

$$u^2 + v^2 = \frac{1}{x^2 + y^2}$$

$$u^2 + v^2 = \frac{1}{3^2} \quad \text{where } r = \frac{1}{3}.$$

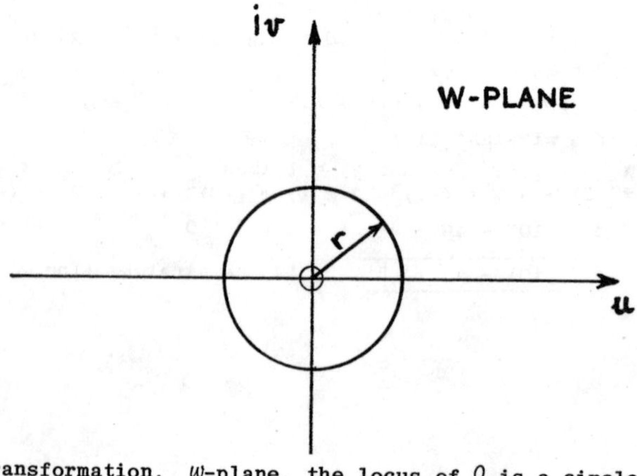

Fig.27 Transformation. w-plane, the locus of Q is a circle $C(0,0)$ and $r = {}^1/_3$

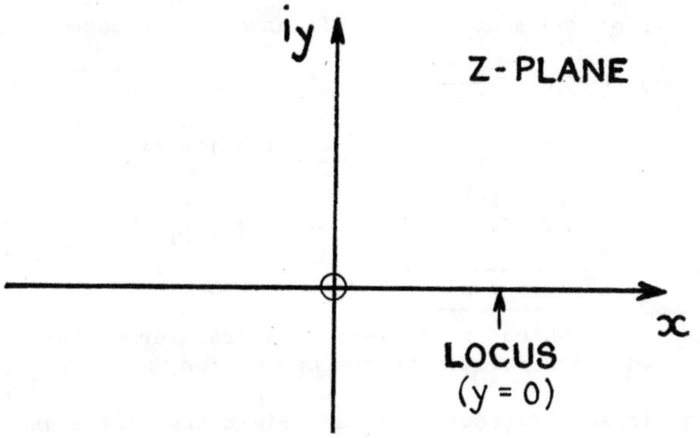

Fig.28 The locus is the x-axis, $y = 0$.

Fig. 27 and Fig. 28 show the paths on the z-plane and the w-plane respectively.

The path on the z-plane is a straight line $x = 3$ and the corresponding path on the w-plane is a circle with a centre at the origin and radius $\frac{1}{3}$.

Therefore, the straight line $x = 3$ displayed on the z-plane is transformed into a circle in the w-plane, if z and w are related by the expression $z = \frac{1}{w}$ and given a condition that $x = 3$ for all values of y.

WORKED EXAMPLE 50

Points P and Q represent the complex numbers $z = x + iy$ and $w = u + iv$ in the z-plane and the w-plane respectively.

Given that z and w are connected by the relation $w = \dfrac{z - i}{z + i}$ and that the locus of P is the x-axis, find the cartesian equation of the locus of Q and sketch the locus of Q on an Argand diagram.

SOLUTIONS 50

Starting with the expression relating z and w, $w = \dfrac{z - i}{z + i}$

then $$w = \frac{x + iy - i}{x + iy + i} = \left[\frac{x + (y - 1)i}{x + (y + 1)i}\right] \cdot \left[\frac{x - (y + 1)i}{x - (y + 1)i}\right]$$

$$w = \frac{\{x + (y - 1)i\}\{x - (y + 1)i\}}{\{x^2 + (y + 1)^2\}} = u + iv.$$

The locus of P is the x-axis, that is, $y = 0$

$$w = \frac{(x - i)(x - i)}{x^2 + 1} = \frac{x^2 + i^2 - 2ix}{x^2 + 1}$$

$$= \frac{x^2 - 1}{x^2 + 1} - 2i\frac{x}{x^2 + 1} = u + iv$$

Equating real and imaginary terms $u = \dfrac{x^2 - 1}{x^2 + 1}$ and $v = \dfrac{-2x}{x^2 + 1}$

(It is required to eliminate x from these equations).

Squaring up both sides of the equations obtain an equation connecting u and v.

$$u^2 = \frac{(x^2 - 1)^2}{(x^2 + 1)^2} \qquad v^2 = \frac{4x^2}{(x^2 + 1)^2}$$

$$\frac{(x^2 - 1)^2 + 4x^2}{(x^2 + 1)^2} = u^2 + v^2 = \frac{x^4 - 2x^2 + 1 + 4x^2}{(x^2 + 1)^2}$$

$$= \frac{x^4 + 2x^2 + 1}{(x^2 + 1)^2} = \frac{(x^2 + 1)^2}{(x^2 + 1)^2} = u^2 + v^2 = 1$$

$$\boxed{u^2 + v^2 = 1}$$

The straight line $\boxed{y = 0}$, which is the x-axis is transformed into a circle on the w-plane if w and z are related by the expression, $w = \frac{z - i}{z + i}$.

The locus of Q is a circle with centre at the origin and radius unity. The z-plane and w-plane loci are shown in Fig. 29 and Fig. 30 respectively.

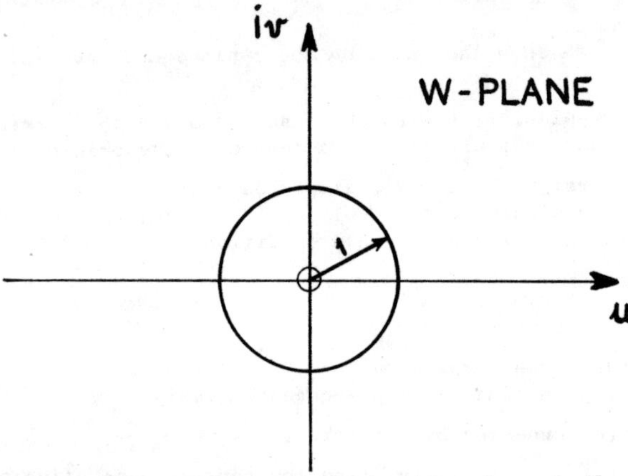

Fig.29 Transformation. The locus is a circle $u^2 + v^2 = 1$ with $c(0,0)$ and $r = 1$.

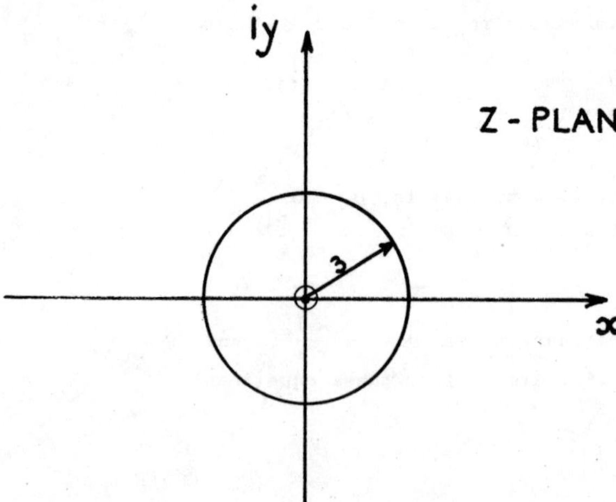

Fig.30 The locus is a circle $c(0,0)$, $r = 3$.

WORKED EXAMPLE 51

Given that $w = \frac{z - i}{z + i}$ find the image in the w-plane of the circle $|z| = 3$ in the z-plane. Illustrate the 2 loci in separate Argand diagrams.

SOLUTION 51

$w = \frac{z - i}{z + i}$ where $z = x + iy$ then

$$w = \frac{x + iy - i}{x + iy + 1} = \frac{x + i(y - 1)}{x + i(y + 1)} \cdot \frac{x - i(y + 1)}{x - i(y + 1)}$$

$$w = \frac{\{x + i(y - 1)\}\{x - i(y + 1)\}}{x^2 + (y + 1)^2}$$

$$w = \frac{x^2 + i(y - 1)x - i(y + 1)x + (y^2 - 1)}{x^2 + y^2 + 2y + 1}$$

$|z| = 3$, $\sqrt{x^2 + y^2} = 3$ since $|x + iy| = \sqrt{x^2 + y^2}$

$x^2 + y^2 = 3^2 = 9$

$$w = \frac{x^2 + y^2 - 1 + i(yx - x - yx - x)}{x^2 + y^2 + 2y + 1}$$

$$w = \frac{x^2 + y^2 - 1 - 2ix}{x^2 + y^2 - 2y + 1} = \frac{9 - 1 - 2ix}{9 + 2y + 1}$$

$$w = \frac{8 - 2xi}{10 + 2y} = \frac{8}{10 + 2y} - i\frac{2x}{10 + 2y} = u + iv$$

Equating real and imaginary terms

$$u = \frac{8}{10 + 2y} \quad \ldots\ldots\ldots\ldots\ldots\ldots \quad (1)$$

$$v = \frac{-2x}{10 + 2y} \quad \ldots\ldots\ldots\ldots\ldots\ldots \quad (2)$$

In order to find the relationship connecting u and v we require to eliminate x and y from (1) and (2)

From equations (1) and (2)

$$10 + 2y = \frac{8}{u} = \frac{-2x}{v}$$

therefore $\frac{u}{v} = \frac{8}{-2x}$ and $\boxed{x = -\frac{4v}{u}}$

$$10 + 2y = \frac{8}{u}, \quad 2y = \frac{8}{u} - 10, \quad y = \left(\frac{4}{u} - 5\right) + \frac{4 - 5u}{u}$$

$$\therefore \quad x^2 + y^2 = 3^2 = \left(-\frac{4v}{u}\right)^2 + \left(\frac{4 - 5u}{u}\right)^2$$

$$\frac{16v^2}{u^2} + \frac{16}{u^2} - \frac{40u}{u} + 25 = 9$$

$$16v^2 + 16 - 40u + 25u^2 - 9u^2 = 0$$

$$16v^2 + 16u^2 - 40u + 16 = 0$$

$v^2 + u^2 - \frac{40}{16}u + 1 = 0$ which is the equation of a circle

$$v^2 + \left(u - \frac{5}{4}\right)^2 - \frac{5^2}{4^2} + 1 = 0 \quad \therefore \quad \boxed{v^2 + \left(u - {}^5/_4\right)^2 = \left(\frac{3}{4}\right)^2}$$

The coordinates of the centre $C\ ({}^5/_4,\ 0)$ and $r = {}^3/_4$ the radius.

The z-plane and w-plane loci are shown in Fig.31 and Fig. 32

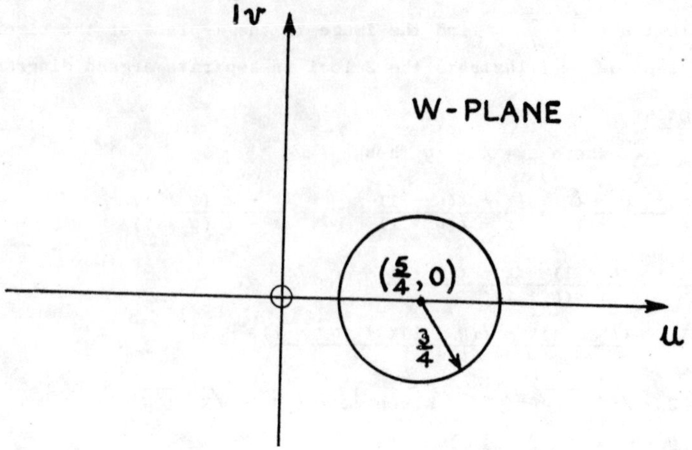

Fig.31 Transformation. The locus is a circle $\left(u - {}^5/_4\right)^2 + v^2 = \left({}^3/_4\right)^2$, $C({}^5/_4, 0)$, $r = {}^3/_4$

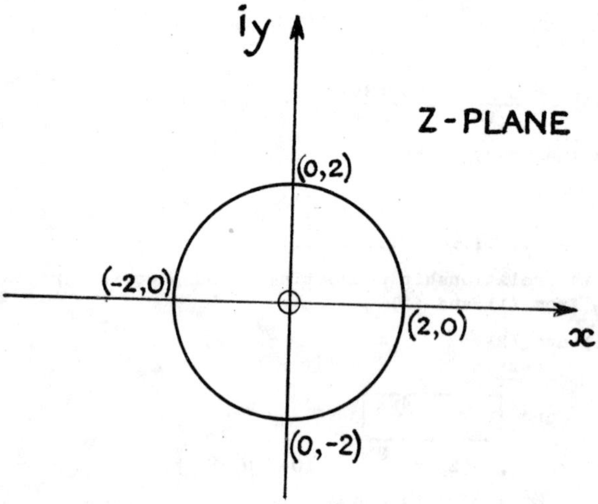

Fig.32 The z-plane is a circle.

WORKED EXAMPLE 52

Given that $w = z + \frac{1}{z}$, find the image in the w-plane of the circle $|z| = 2$ in the z-plane. Illustrate the 2 loci in separate Argand diagrams.

SOLUTION 52

$|z| = 2$ is a circle with centre the origin and radius 2.

If $z = x + iy$, then $|x + iy| = \sqrt{x^2 + y^2} = 2$

or $\boxed{x^2 + y^2 = 2^2}$

The locus is illustrated in the Argand diagram of Fig. 33

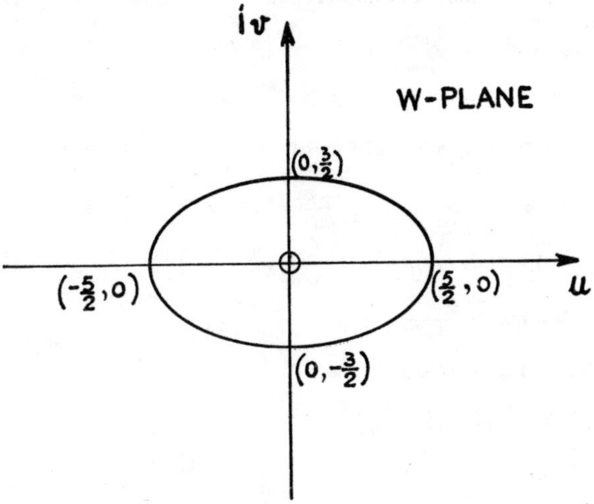

Fig.33 Transformation. The W-plane is an ellipse.

$$w = x + iy + \frac{1}{x + iy} = x + iy + \frac{x - iy}{x^2 + y^2} = x + \frac{x}{x^2 + y^2} + i\left(y - \frac{y}{x^2 + y^2}\right)$$

$$w = u + iv = x + \frac{x}{x^2 + y^2} + i\left(y - \frac{y}{x^2 + y^2}\right).$$

Equating real and imaginary terms

$u = x + \dfrac{x}{x^2 + y^2}$ and $v = y - \dfrac{y}{x^2 + y^2}$

Since $x^2 + y^2 = 4$, $u = x + \dfrac{x}{4}$ and $v = y - \dfrac{y}{4}$

$u = \dfrac{5x}{4}$ $v = \dfrac{3y}{4}$ $x = \dfrac{4u}{5}$ and $y = \dfrac{4v}{3}$

Squaring up both of these quantities

$$x^2 + y^2 = 4 = \left(\frac{4u}{5}\right)^2 + \left(\frac{4v}{3}\right)^2$$

Therefore, $\dfrac{u^2}{\left(5/4\right)^2} + \dfrac{v^2}{\left(3/4\right)^2} = 1$ the locus in the W-plane.

The circle in the Z-plane has a radius of 2 and the centre is C (0,0), this is transformed to the W-plane as an ellipse.
Fig. 34 illustrates this point.

Fig.34 The locus is a circle from A to B to C.

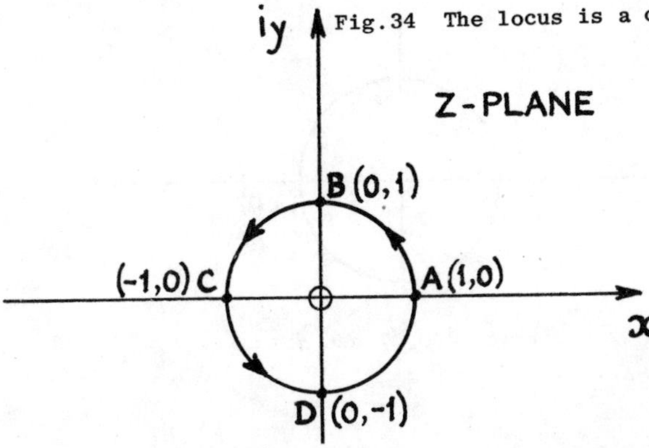

WORKED EXAMPLE 53

Find the image on the W-plane of the circles (i) $|Z| = 1$ and (ii) $|Z| = 3$ under the function $W = Z + \frac{1}{Z}$.

SOLUTION 53

(i) $|Z| = 1$ or $x^2 + y^2 = 1$ a circle centred at the origin with unity radius

$$W = Z + \frac{1}{Z} = x + iy + \frac{1}{x + iy} = x + iy + \frac{x - iy}{x^2 + y^2}$$

$$= \left(x + \frac{x}{x^2 + y^2}\right) + i\left(y - \frac{y}{x^2 + y^2}\right)$$

$$W = u + iv = x + \frac{x}{x^2 + y^2} + i\left(y - \frac{y}{x^2 + y^2}\right)$$

Equating real and imaginary terms

$$u = x + \frac{x}{x^2 + y^2} = x + \frac{x}{1} = 2x$$

$$v = y - \frac{y}{x^2 + y^2} = y - \frac{y}{1} = 0$$

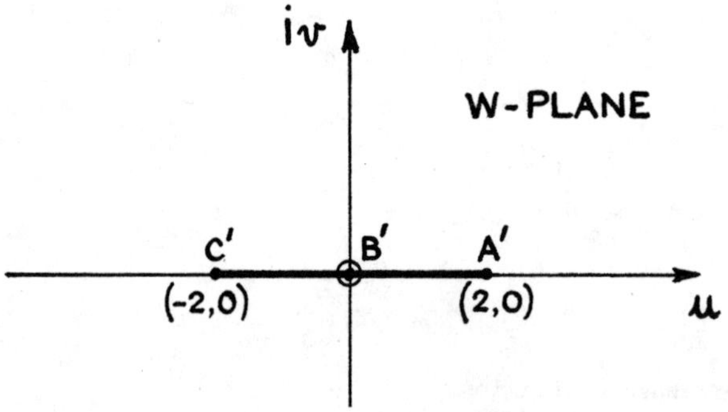

Fig.35 Transformation. The locus is a straight line from A' to B' to C'.

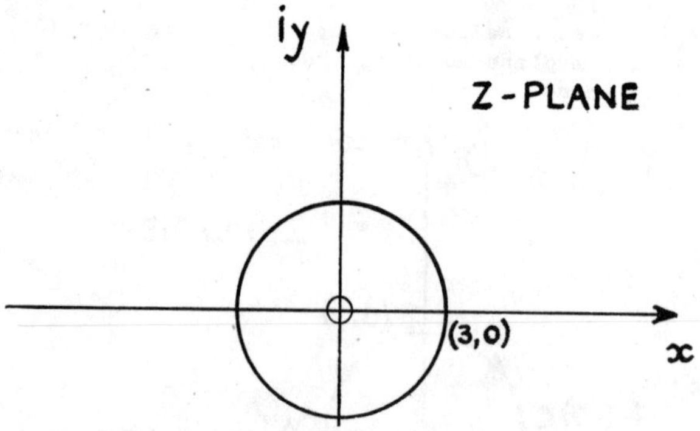

Fig.36 The locus is a circle, $c(0,0)$ and $r = 3$.

Referring to Fig. 35 and Fig. 36 where in the z - plane the circle is transformed to the straight line of the W - plane. When A moves to B , that is, $u = 2$ when $x = 1$ and $u = 0$ when $x = 0$ at B, then from B to C, that is when $x = 0$, $u = 0$, and when $x = -1$, $u = -2$ then it moves from C to D, that is, when $x = 0$, $u = 0$ and when $u = 2$, $x = 1$. Therefore, Z travels round the circle once, from A back to A in anticlockwise direction, while W travels from A' to C' via B' and back again to A' via B'.

(ii) $|Z| = 3$ or $x^2 + y^2 = 3$ a circle centred at the origin with radius 3.

$$W = Z + \frac{1}{Z} = x + iy + \frac{1}{x + iy} = x + iy + \frac{x - iy}{x^2 + y^2}$$

$$= \left(x + \frac{x}{x^2 + y^2}\right) + i\left(y - \frac{y}{x^2 + y^2}\right)$$

$$W = u + iv = x + \frac{x}{x^2 + y^2} + i\left(y - \frac{y}{x^2 + y^2}\right)$$ Equating real & imaginary ter

$$u = x + \frac{x}{x^2 + y^2} \quad \text{and} \quad v = y - \frac{y}{x^2 + y^2}$$

$$u = x + \frac{1}{9}x \quad \text{and} \quad v = y - \frac{1}{9}y = \frac{8}{9}y$$

$$u = \frac{10}{9}x \quad \text{therefore} \quad y = \frac{9v}{8} \quad \text{and} \quad x = \frac{9u}{10}$$

$$x^2 + y^2 = \left(\frac{9v}{8}\right)^2 + \left(\frac{9u}{10}\right)^2 = 9. \quad \text{Therefore} \quad \frac{v^2}{\left(\frac{8}{9}\right)^2} \quad \frac{u^2}{\left(\frac{10}{9}\right)^2} = 1$$

The transformation is illustrated in Fig. 37 and Fig. 38

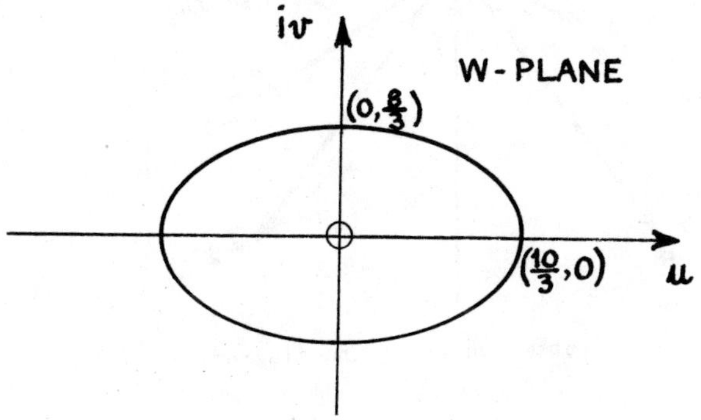

Fig.37 The transformation is illustrated on the w-plane.
It is an ellipse $\frac{x^2}{(10/3)^2} + \frac{y^2}{(8/3)^2} = 1$

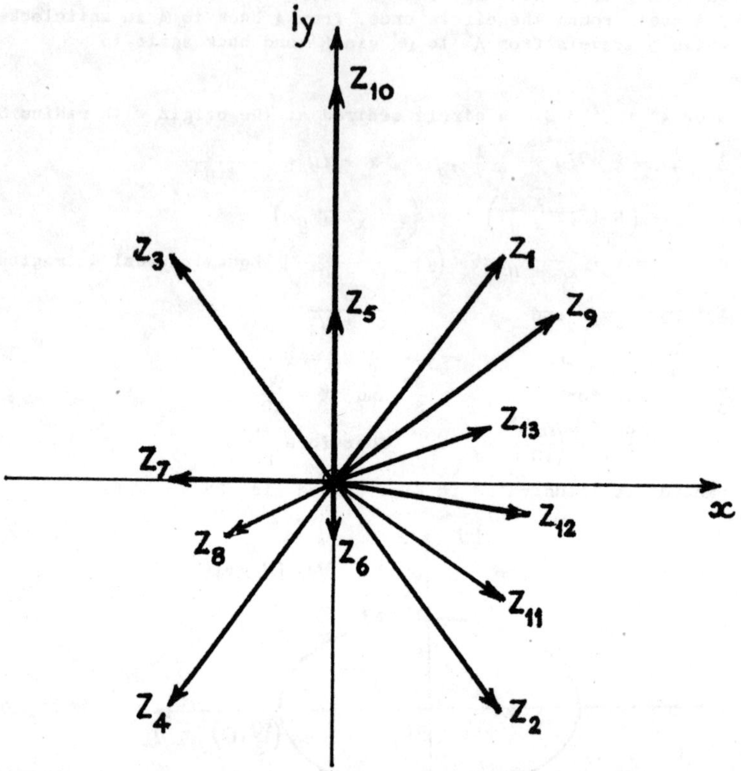

Fig.38 Complex numbers plotted on an Argand diagram.

WORKED EXAMPLE 54

If Z and W represent points P and Q in the Argand diagram and $|Z| = 1$, arg Z steadily increases from $-\pi$ to $+\pi$, describe the corresponding motion of Q if $W = Z^{1/3}$.

SOLUTIONS 54

$Z = \cos\theta + i\sin\theta$ and $Z^{1/3} = (\cos\theta + i\sin\theta)^{1/3}$

$Z^{1/3} = \cos\dfrac{\theta}{3} + i\sin\dfrac{\theta}{3}$ or $\cos\dfrac{\theta+2\pi}{3} + i\sin\dfrac{\theta+2\pi}{3}$

or $\cos\dfrac{\theta-2\pi}{3} + i\sin\dfrac{\theta-2\pi}{3}$.

For each position of P, there are 3 positions of Q (Q_1, Q_2, Q_3) which move continuously along the circle $|Z|$, anti-clockwise.

Q_1 moves from $\theta = -\dfrac{\pi}{3}$ to $\theta = +\dfrac{\pi}{3}$, and at the same time,

Q_2 moves from $\theta = \dfrac{\pi}{3}$ to $\theta =$ and Q_3 moves from $\theta = -\pi$ to $\theta = -\pi/3$.

EXERCISES 20

If Z and W represent complex numbers of two points P and Q respectively, and $|Z| = 1$, P moves so that arg Z steadily increases from $-\pi$ to π.

Describe the corresponding motion of Q when

1. $W = 2z + 3$ (Ans.   circle $c(3,0)$ $r = 2$)

2. $W = 2 + iz$ (Ans.   circle $c(2,0)$ $r = 1$)

3. $W = 3z^2$ (Ans. $|Z| = 3$, twice circle)

4. $W = Z^3$ (Ans. $|Z| = 1$, 3 times circle)

5. $W = Z^{-\frac{1}{2}}$ (Ans.   Two semi-circles of $|Z| = 1$)

6. $W = Z^2 + 2z$ (Ans.   Cardioid $r = 2(1 + \cos\theta)$ displaced 1 unit)

7. $W = (Z + 1)^{\frac{1}{2}}$ (Ans.   Right loop of lemriscate $r^2 = 2\cos 2\theta$ to the left).

PART II

SOLUTIONS OF ALL THE EXERCISES SET

SOLUTIONS 1

1\. (i) $\sqrt{-2} = \sqrt{-1}\,\sqrt{2} = \sqrt{2}i$ where $\sqrt{-1} = i$

(ii) $\sqrt{-4} = \sqrt{-1}\,\sqrt{4} = 2i$

(iii) $\sqrt{-8} = \sqrt{-1}\,\sqrt{8} = 2\sqrt{2}i$

(iv) $\sqrt{-16} = \sqrt{-1}\,\sqrt{16} = 4i$

(v) $\sqrt{-27} = \sqrt{-1}\,\sqrt{27} = 3\sqrt{3}i$

The above complex numbers are wholly imaginary.

(vi) $1 + \sqrt{-3} = 1 + \sqrt{-1}\,\sqrt{3} = 1 + i\sqrt{3}$

(vii) $-1 - \sqrt{-5} = -1 - \sqrt{-1}\,\sqrt{5} = -1 - i\sqrt{5}$

(viii) $-5 + \sqrt{-7} = -5 + \sqrt{-1}\,\sqrt{7} = -5 + i\sqrt{7}$

The above complex numbers are expressed as real and imaginary.

2\. (i) $3x^2 - x + 1 = 0$
The discriminant $D = b^2 - 4ac = (-1)^2 - 4 \times 3 \times 1 = -11$ which is negative, therefore the roots are complex.

(ii) $-x^2 + x - 5 = 0$
The discriminant $D = (1)^2 - 4(-5)(-1) = -19$ which is also negative and the roots are complex.

(iii) $-5x^2 + 7x + 5 = 0$
$D = (7)^2 - 4(-5)(5) = 149$. This is positive and the roots are real.

(iv) $x^2 - 4x + 8 = 0$
The discrimant, $D = (-4)^2 - 4(8) = -16$. The roots are complex.

(v) $x^2 + 2x + 2 = 0$
$D = 4 - 2(4) = -4$. The roots are complex.

3\. (i) $3x^2 - x + 1 = 0$

$x = \dfrac{1 \pm \sqrt{-11}}{6} = \dfrac{1}{6} \pm i\sqrt{11}/6$ $\quad x_1 = \dfrac{1}{6} + i\sqrt{11/6}$ and $x_2 = 1/6 - i\sqrt{11/6}$

(ii) $-x^2 + x - 5 = 0$

$x = \dfrac{-1 \pm \sqrt{-19}}{-2}$; $x_1 = \dfrac{1}{2} - \sqrt{19}i/2$ and $x_2 = \dfrac{-1 - i\sqrt{19}}{-2} = \dfrac{1}{2} + i\sqrt{19}/2$

(iii) $x^2 - 4x + 8 = 0$ $\quad x = \dfrac{4 \pm \sqrt{-16}}{2} = 2 \pm i2$, $x_1 = 2 + i2$, $x_2 = 2 - i2$

(iv) $x^2 + 2x + 2 = 0$, $\quad x_1 = -1 + i$ and $x_2 = -1 - i$
The roots appear in conjugate pairs.

4 (i) Solving the simultaneous equations

$$x^2 + y^2 = 1, \quad \text{and} \quad 3x - y + 1 = 0,$$
$$x^2 + (3x + 1)^2 - 1 = 0$$
$$x^2 + 9x^2 + 6x + 1 - 1 = 0$$
$$10x^2 + 6x = 0$$

Factorising $2x(5x + 3) = 0$ and solving gives $x = 0$ and $x = -3/5$.
Substituting for y gives the corresponding values $x = 1$ and $y = -4/5$.

The graphs intersect at (0,1), and (-3/5, -4/5).

(ii) Solving the simultaneous equations $x^2 = 4y$ and $-x^2 = 4y$ gives $x = 0$ and $y = 0$.
The graphs intersect at the origin (0,0), since there is only one point, the graphs touch each other.

(iii) Solving the simultaneous equations $x^2 = 4y$ and $x - y = 3$

$$(3 + y)^2 = 4y \text{ or } y^2 + 6y + 9 - 4y = 0$$

therefore $y^2 + 2y + 9 = 0$

which gives

$$y = \frac{-2 \pm \sqrt{4 - 36}}{2} = \frac{-2 \pm \sqrt{-32}}{2} = -1 \pm i2\sqrt{2}$$

hence $y_1 = -1 + 2\sqrt{2}i$ and $y_2 = -1 - 2\sqrt{2}i$

The roots are complex and the graphs do not intersect.

(iv) Solving the simultaneous equations

$$x^2 + (y - 1)^2 = 1 \text{ and } y + 3x = 4$$
$$x^2 + (4 - 3x - 1)^2 = 1, \text{ or } x^2 + 9 + 9x^2 - 18x - 1 = 0$$
$$10x^2 - 18x + 8 = 0, \text{ or } 5x^2 - 9x + 4 = 0$$

giving $x_1 = 1$ and $x_2 = 4/5$

The graphs intersect at (1, 1) and (4/5, 8/5)

SOLUTIONS 2

1. (i) $1 + 3i$ (ii) $2 + 5i$ (iii) $0 + 6i$ (iv) $3 + 0i$ (v) $-1 + 3i$ (vi) $2 - 4i$ (vii) $0 + 0i$ (viii) $a + bi$ (ix) $x + yi$ (x) $-3 - 4i$

2. (i) (3,4) (ii) (3, -4) (iii) (-3, 4) (iv) (-3, -4) (v) (0,3) (vi) (0, -1) (vii) (-3,0) (viii) (-2, -1) (ix) (b, a) (x) (0,7) (xi) (3, -2) (xii) $(x, -y)$ (xiii) $(\cos\theta, \sin\theta)$

3.

4.

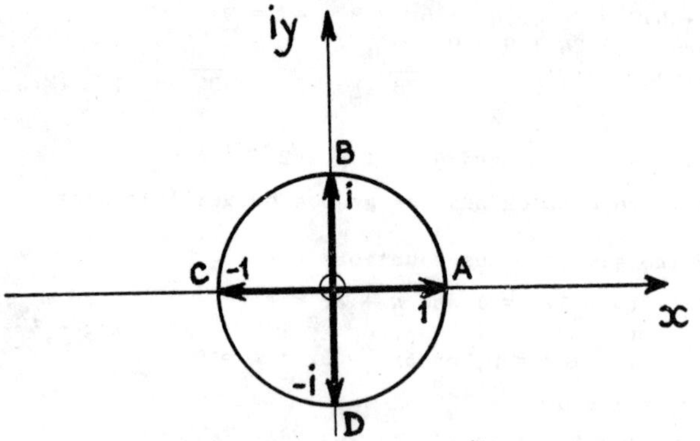

Fig.1 The powers of i, $i^{1986} = -1$.

If OA represents one unity along the positive x-axis of the Argand diagram, and OA is to be rotated in an anti-clockwise direction of 90°, as shown in Fig. 1 in position OA, this means that $OA \times i = OB$

Multiplying a vector by i, the vector is rotated through 90° in an anticlockwise direction. If the vector OB is rotated further through 90° in anti-clockwise direction, in position OC, this means that $OB \times i = OC = (-1)$.

The vector OC is rotated through 90° further, in position OD, i.e. $OD = OC \times i = -(i) = -i$

and finally, $OD \times i = OA$ thus completing one revolution.

(i) $i^2 = -1$ (ii) $i^5 = i$ (iii) $i^7 = -i$ (iv) $i^{33} = i$ (v) $i^{1986} = -1$.

The vector i^{1986} completes 496 revolutions and two quarters of a revolution and hence is in the position OC which is equal to -1.

5. A complex number is a vector.
A vector has magnitude and direction.
Referring to Fig. 1, OA is of unity magnitude and is in the direction of zero degrees, the vector OB is of unity magnitude and is in the direction of the positive y-axis i.e. 90° in an anticlockwise direction.

$Z = 3i$, means that there are three units in the direction i, the positive y-axis.
$Z = -5i$, means that there are five units in the direction $-i$, the negative y-axis.
$Z = -7$, means that there are seven units in the direction of the negative x-axis.

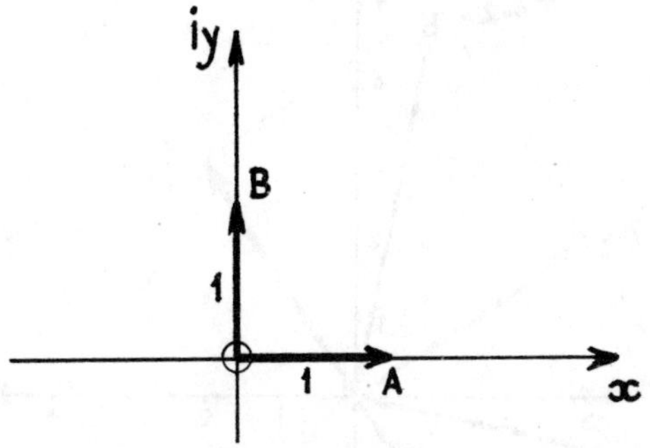

Fig.2 Vectors $OA = 1 \angle 0^\circ$, $OB = 1 \angle 90^\circ$.

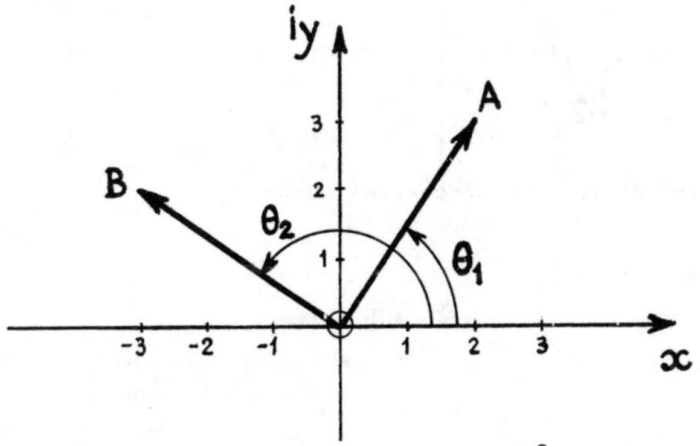

Fig.3 OA is perpendicular to OB, $\theta_2 - \theta_1 = 90^\circ$.

SOLUTIONS 3

1. (i) $Z_1 + Z_2 = (2 + 3i) + (3 + 4i) = 5 + 7i$
   (ii) $Z_1 + Z_3 = (2 + 3i) + (-4 - i5) = -2 - i2$
   (iii) $Z_2 + Z_3 = (3 + 4i) + (-4 -5i) = -1 - i1$
   (iv) $Z_1 - Z_2 = (2 + 3i) - (3 + 4i) = -1 - i1$
   (v) $Z_1 - Z_3 = (2 + 3i) - (-4 - 5i) = 6 + i8$
   (vi) $Z_3 - Z_2 = (-4 - i5) - (3 + 4i) = -7 - i9$
   (vii) $2Z_1 + 3Z_3 = 2(2 + 3i) + 3(-4 - i5) = 4 + 6i - 12 - i15 = -8 - i9$
   (viii) $Z_1 + 2Z_2 = (2 + 3i) + 2(3 + 4i) = 2 + 3i + 6 + 8i = 8 + 11i$
   (ix) $Z_3 - 3Z_1 = (-4 - i5) - 3(2 + 3i) = -4 - i5 - 6 - 9i = -10 - 14i$
   (x) $3Z_3 - 2Z_1 = 3(-4 - i5) - 2(2 + 3i) = -12 - i15 - 4 - 6i = -16 - 21i$
   (xi) $Z_3 + 5Z_2 = (-4 - i5) + 5(3 + 4i) = -4 - i5 + 15 + 20i = 11 + 15i$

2. (1) $Z = x + iy$
(ii) $Z = -3 + i5$
(iii) $Z = a - ib$

3. $E_1 + E_2 = (20 + i30) + (10 + i15) = 30 + i45$

$E_1 - E_2 = (20 + i30) - (10 + i15) = 10 + i15$

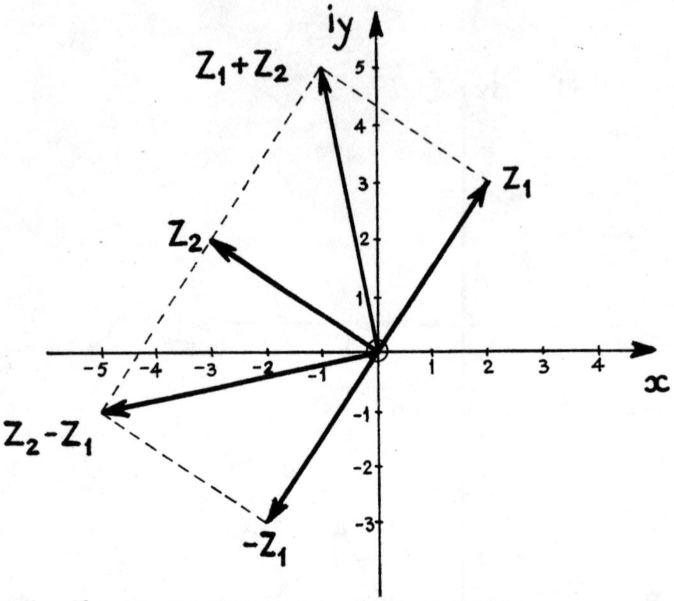

Fig.4 The sum and difference of complex numbers.

4. Refer to Fig. 4

$\tan \theta_1 = 3/2 = m_1$

$\tan \theta_2 = -2/3 = m_2$

$$\tan A\hat{O}B = \tan(\theta_2 - \theta_1) = \frac{\tan\theta_2 - \tan\theta_1}{1 + \tan\theta_1 \tan\theta_2} = \frac{-2/3 - 3/2}{1 + (3/2)(-2/3)}$$

$$= \frac{-2/3 - 3/2}{1 - 1} = \frac{-13/6}{0} = \infty$$

$m_1 m_2 + 1 = 0$, from which we deduce that $m_1 m_2 = -1$ and therefore the vectors OA and OB are perpendicular and $\boxed{A\hat{O}B = \pi/2}$

5. $Z_1 = -3 + i2$
$Z_2 = 2 + i3$

The resultant $Z_1 + Z_2 = (-3 + i2) + (2 + i3) = -1 + i5$
$Z_2 - Z_1 = (-3 + i2) - (2 + i3) = -5 - i1$

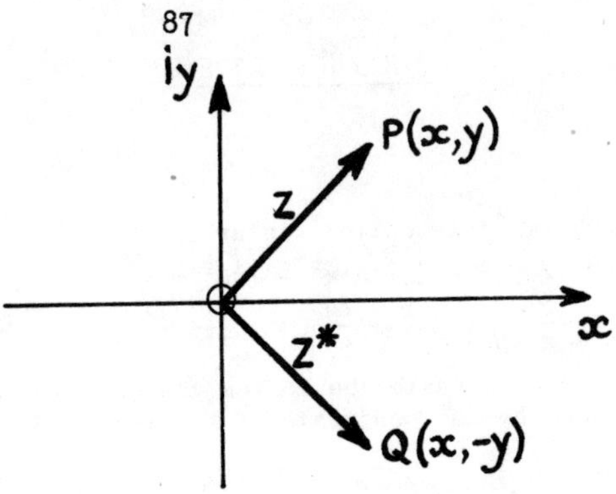

Fig.5   Conjugate of z is z^*.

SOLUTIONS 4

1. (i) $Z_1Z_2 = (3 - 4i)(1 + i) = 3 - 4i + 3i - 4i^2 = 3 - i - 4(-1) = 7 - i$

    (ii) $Z_1Z_3 = (3 - 4i)(2 + 3i) = 6 - 8i + 9i - 12i^2 = 18 + i$

    (iii) $Z_2Z_3 = (1 + i)(2 + 3i) = 2 + 2i + 3i + 3i^2 = -1 + 5i$

    (iv) $Z_1Z_2Z_3 = (Z_1Z_2)\, Z_3 = (7 - i)(2 + 3i) = 14 - i2 + 21i - 3i^2 = 17 + 19i$

           $= Z_1(Z_2Z_3) = (3 - 4i)(-1 + 5i) = -3 + 4i + 15i + 20 = 17 + 19i$

           $= Z_2(Z_1Z_3) = (1 + i)(18 + i) = 18 + 18i + i + i^2 \qquad = 17 + 19i$

2. (i) $(3i)(5i) = 15i^2 = -15$

    (ii) $(2 + 3i)(3 + 4i) = 6 + 9i + 8i + 12i^2 = -6 + 17i$

    (iii) $(3 - 5i)(3 + 4i) = 9 - 15i + 12i - 20i^2 = 29 - 3i$

    (iv) $(4 - 5i)(1 + i) = 4 - 5i + 4i - 5i^2 = 9 - i$

    (v) $(1 + 2i)^3 = 1 + 3(2i) + 3(2i)^2 + (2i)^3$
           $= 1 + 6i - 12 + 8i^3 = -11 + 6i - 8i = -11 - 2i$

    (vi) $5i(1 - i) = 5i - 5i^2 = 5 + 5i$

    (vii) $(1 + i)(1 - i) = 1 - i^2 = 2 + i0$ $(viii)(1+2i)(1-2i) = 5 + i0.$

    (ix) $(1 - 3i)(1 + 3i) = 1 - 9i^2 = 10 + i0$

    (x) $(4 + 3i)^2 = 16 + 24i + 9i^2 = 7 + 24i$

    (xi) $(a + bi)^2 = a^2 + 2abi + b^2i^2 = (a^2 - b^2) + 2abi$

    (xii) $(\cos \theta + i \sin \theta)(\cos \phi + i \sin \phi)$
           $= \cos \theta \cos \phi + i \sin \theta \cos \phi + i \cos \theta \sin \phi - \sin \theta \sin \phi$
           $= (\cos \theta \cos \phi - \sin \theta \sin \phi) + i(\sin \theta \cos \phi + \cos \theta \sin \phi)$
           $= \cos (\theta + \phi) + i \sin (\theta + \phi)$

    (xiii) $(1 + 3i)^3 = 1 + 3 \times 3i + 3(3i)^2 + (3i)^3 = 1 + 9i + 27i^2 + 27i^3$
           $= 1 + 9i - 27 - 27i = -26 - 18i$

    (xiv) $(1 - i)^3 = 1 - 3i + 3i^2 - i^3 = 1 - 3i - 3 + i = -2 - 2i$

    (xv) $(1 - i^2)^5 = (1 + 1)^5 = 2^5 = 32 = 32 + 0i$

3. If $Re(Z_1Z_2) = x_1x_2 - y_1y_2$

        $Im(Z_1Z_2) = x_1y_2 + y_1x_2$

        then $Z_1Z_2 = (x_1x_2 - y_1y_2) + i(x_1y_2 + y_1x_2)$

CONJUGATES

SOLUTIONS 5/6

1. $Z = x + iy$
$Z^* = x - iy$ where Z^* is called Z star

(i) $ZZ^* = (x + iy)(x - iy) = x^2 - i^2y^2 = x^2 + y^2$ $\boxed{ZZ^* = x^2 + y^2}$

(ii) $\frac{1}{Z} = \frac{1}{x + iy} \cdot \frac{x - iy}{x - iy} = \frac{x - iy}{x^2 - i^2y^2} = \frac{x}{x^2 + y^2} - i\frac{y}{x^2 + y^2}$.

In order to eliminate the i from the denominator, $\frac{1}{x + iy}$ is multiplied and divided by the conjugate of $x + iy$.

therefore $\frac{1}{Z} = \frac{x}{x^2 + y^2} - i\frac{y}{x^2 + y^2}$ and its conjugate

$\left(\frac{1}{Z}\right)^* = \frac{x}{x^2 + y^2} + i\frac{y}{x^2 + y^2} = \frac{x + iy}{x^2 + y^2}$

$= \frac{Z}{ZZ^*} = \frac{1}{Z^*}$, where ZZ^* is $x^2 + y^2$ and $Z = x + iy$

Therefore $\boxed{\left(\frac{1}{Z}\right)^* = \frac{1}{Z^*}}$

2. If $Z = x + iy$ where $x, y, \in \mathbb{R}$
then the reflection in the x-axis of the point P is Q with coordinates $Q(x, -y)$ then the conjugate $Z^* = x - iy$

$Z_1 = x_1 + iy_1$ $\qquad Z_1^* = x_1 - iy_1$

$Z_2 = x_2 + iy_2$ $\qquad Z_2^* = x_2 - iy_2$

$Z_1 + Z_2 = x_1 + iy_1 + x_2 + iy_2 = (x_1 + x_2) + i(y_1 + y_2)$
$(Z_1 + Z_2)^* = (x_1 + x_2) - i(y_1 + y_2) = (x_1 - iy_1) + (x_2 - iy_2)$

$\boxed{(Z_1 + Z_2)^* = Z_1^* + Z_2^*}$

3. If $\frac{1}{Z} = \frac{iy + x}{iy - x}$

$\frac{1 + Z^2}{2Z} = \frac{1}{2Z} + \frac{Z}{2} = \frac{1}{2}\left(\frac{iy + x}{iy - x}\right) + \frac{1}{2}\cdot\left(\frac{iy - x}{iy + x}\right)$

$= \frac{1}{2}\left[\frac{(iy + x)(iy + x) + (iy - x)(iy - x)}{i^2y^2 - x^2}\right]$

where $(iy - x)(iy + x) = i^2y^2 - x^2$, the difference of squares

$\frac{1 + Z^2}{2Z} = \frac{1}{2}\left[\frac{i^2y^2 + x^2 + 2iyx + i^2y^2 - 2iyx + x^2}{i^2y^2 - x^2}\right] = \frac{1}{2}\left[\frac{2i^2y^2 + 2x^2}{i^2y^2 - x^2}\right]$

$= \frac{1}{2}\left(\frac{2x^2 - 2y^2}{-y^2 - x^2}\right) = -\frac{x^2 - y^2}{x^2 + y^2} = \frac{y^2 - x^2}{y^2 + x^2}$

therefore $\boxed{\frac{2Z}{1 + Z^2} = \frac{x^2 + y^2}{y^2 - x^2}}$ where $y \neq x$.

This can be written as $\frac{y^2 - x^2}{y^2 + x^2} = \frac{1 + Z^2}{2Z}$ so that $y = x$ can be defined.

4. $$\frac{z}{1+z^2} = \frac{x+iy}{1+(x+iy)^2} = \frac{x+iy}{1+x^2+i^2y^2+2ixy} = \frac{x+iy}{1+x^2-y^2+i2xy}$$

Multiplying numerator and denominator by the conjugate of $1 + x^2 - y^2 + i2xy$, we have

$$\frac{z}{1+z^2} = \frac{(x+iy)(1+x^2-y^2-i2xy)}{\{1+x^2-y^2+i2xy\}\{1+x^2-y^2-i2xy\}}$$

$$\{(1+x^2-y^2)+i2xy\}\{(1+x^2-y^2)-i2xy\}$$
$$=\{(1+x^2-y^2)^2-i^2 4x^2y^2\},$$

$$\frac{z}{1+z^2} = \frac{(x+iy)(1+x^2-y^2-i2xy)}{(1+x^2-y^2)^2+4x^2y^2}$$

$$= \frac{x(1+x^2-y^2) - i^2\,2xy^2 - i2x^2y + iy(1+x^2-y^2)}{(1+x^2-y^2)^2+4x^2y^2}$$

$\frac{z}{1+z^2} = Re\left(\frac{z}{1+z^2}\right) + iIm\left(\frac{z}{1+z^2}\right)$ where $Re\left[\frac{z}{1+z^2}\right]$ is the real part

$$Re\left[\frac{z}{1+z^2}\right] = \frac{x(1+x^2-y^2)+2xy^2}{(1+x^2-y^2)^2+4x^2y^2} = \frac{x+x^3-xy^2+2xy^2}{(1+x^2-y^2)^2+4x^2y^2}$$

$$= \frac{x+x^3+xy^2}{1+x^4+y^4+2x^2-2y^2-2x^2y^2+4x^2y^2}$$

$$= \frac{x(1+x^2+y^2)}{1+x^4+y^4+2x^2-2y^2+2x^2y^2}$$

In order that $\frac{z}{1+z^2}$ is real, then the imaginary part

$$Im\left[\frac{z}{1+z^2}\right] = 0$$

$$Im\left[\frac{z}{1+z^2}\right] = \frac{y(1+x^2-y^2)-2x^2y}{(1+x^2-y^2)^2+4x^2y^2} = 0$$

therefore $y(1+x^2-y^2) = 2x^2y$ dividing both sides by y where $y \neq 0$, then $1 + x^2 - y^2 = 2x^2$

therefore $\boxed{x^2 + y^2 = 1}$ which is a circle with centre the origin(0,0) and radius 1.

$$\frac{z}{1+z^2} = \frac{(1+x^2+y^2)x}{1+x^4+y^4+2x^2-2y^2+2x^2y^2} + iy\frac{(1-x^2-y^2)}{(1+x^2-y^2)+4x^2y^2}$$

where $a = \dfrac{(1+x^2+y^2)x}{1+x^4+y^4+2x^2-2y^2+2x^2y^2}$

and $b = \dfrac{(1-x^2-y^2)y}{(1+x^2-y^2)^2+4x^2y^2}$

5. $$\frac{1}{Z} = \frac{1}{Z_1} + \frac{1}{Z_2} = \frac{Z_1+Z_2}{Z_1Z_2}$$

$$Z = \frac{Z_1Z_2}{Z_1+Z_2} = \frac{(1+i)(1-2i)}{1+i+1-2i} = \frac{1+i-2i+2}{2-i} = \frac{3-i}{2-i}$$

Multiplying numerator and denominator by the conjugate of $2 - i$, which is $2 + i$

$$Z = \frac{3-i}{2-i} \times \frac{2+i}{2+i} = \frac{6-2i+3i-i^2}{4+1} = \frac{6+i+1}{5} = \frac{7}{5}+i\frac{1}{5}$$

where $a = 7/5$ and $b = 1/5$.

6. $$\frac{1}{u+iv} = 3+4i$$

$$\frac{1}{3+4i} = u+iv = \frac{3-4i}{(3-4i)(3+4i)} = \frac{3-4i}{9+16} = \frac{3}{25} - \frac{4}{25}i$$

Multiplying numerator and denominator by the conjugate of $3 + 4i$, which is $3 - 4i$.

$$u = \frac{3}{25} \text{ and } v = -\frac{4}{25}$$

Note that $(3 - 4i)(3 + 4i)$ is always a positive real quantity because $3^2 - (4i)^2 = 9 - 16i^2 = 9 - 16(-1) = 9 + 16 = 25$.

7. $$\frac{1}{x + iy} = 5 - 12i$$

Multiplying numerator and denominator by the conjugate of $5 - 12i$

$$\frac{1}{5 - 12i} = x + iy = \frac{1}{5 - 12i} \cdot \left(\frac{5 + 12i}{5 + 12i}\right) = \frac{5 + 12i}{25 - 12^2 i^2} = \frac{5 + 12i}{25 + 144}$$

$$x + iy = \frac{5}{169} + i\frac{12}{169}$$

therefore $x = \frac{5}{169}$, $y = \frac{12}{169}$.

8. $$Z = \frac{3 + 4i}{5 - 12i} \cdot \left(\frac{5 + 12i}{5 + 12i}\right) = \frac{15 + 20i + 36i + 48i^2}{5^2 - 12^2 i^2} = \frac{(15 - 48) + 56i}{25 + 144}$$

$$= \frac{-33}{169} + \frac{56i}{169}$$

Multiplying numerator and denominator by the conjugate of $5 - 12i$ as above, then $Z = \frac{-33}{169} + \frac{56}{169}i$.

9. (i) $$(1 - i)^{-2} + (1 + i)^{-2} = \frac{1}{(1 - i)^2} + \frac{1}{(1 + i)^2} = \frac{1}{1 - 2i + i^2} + \frac{1}{1 + 2i + i^2}$$

$$= \frac{1}{1 - 2i - 1} + \frac{1}{1 + 2i - 1} = -\frac{1}{2i} + \frac{1}{2i} = 0$$

therefore $(1 - i)^{-2} + (1 + i)^{-2} = 0$

(ii) $$(1 - i)^{-3} + (1 + i)^{-3} = \frac{1}{(1 - i)^3} + \frac{1}{(1 + i)^3}$$

$$= \frac{1}{1 - 3i + 3i^2 - i^3} + \frac{1}{1 + 3i + 3i^2 + i^3}$$

$$= \frac{1}{1 - 3i - 3 + i} + \frac{1}{1 + 3i - 3 - i}$$

$$= \frac{1}{-2 - 2i} + \frac{1}{-2 + 2i} = \frac{-2 + 2i - 2 - 2i}{(-2 - 2i)(-2 + 2i)}$$

$$= \frac{-4}{-(2i + 2)(2i - 2)} = -\frac{4}{4i^2 - 4} = \frac{4}{8} = \frac{1}{2}$$

and $(1 - i)^{-3} + (1 + i)^{-3} = \frac{1}{2}$

(iii) $$(1 - i)^{-4} + (1 + i)^{-4} = \frac{1}{(1 - i)^4} + \frac{1}{(1 + i)^4} = \frac{1}{(1 - i)^2(1 - i)^2} + \frac{1}{(1 + i)^2(1 + i)^2}$$

$$= \frac{1}{(1 - 2i + i^2)^2} + \frac{1}{(1 + 2i + i^2)^2} = \frac{1}{4i^2} + \frac{1}{4i^2}$$

$$(1 - i)^{-4} + (1 + i)^{-4} = -\frac{1}{4} - \frac{1}{4} = -\frac{1}{2}$$

We could have expanded $(1 - i)^4$ and $(1 + i)^4$ by using the binomial theorem, but it is easier to do as above.

$$(1 - i)^{-4} + (1 + i)^{-4} = -\frac{1}{2}$$

10. $(2 + i)x + (1 + 3i)y + 2 = 0$
$2x + ix + y + 3iy + 2 = 0$

Equating real and imaginary terms

$2x + y + 2 = 0$ ....(1)

$x + 3y = 0$.......................... (2)

Solving the simultaneous equations

from (2) $x = -3y$, and substituting in (1)

$2(-3y) + y + 2 = 0$

$-5y = -2$

$\boxed{y = +\frac{2}{5}}$

$x = (-3)(\frac{2}{5})$

$\boxed{x = -\frac{6}{5}}$

11. $Z^2 - 6Z + 25 = 0$ ....(1)

If $Z = x + iy$ or $Z = 3 + i4$, substituting this equation in (1)

$(3 + i4)^2 - 6(3 + i4) + 25 = 0$

$9 + 24i + 16i^2 - 18 - i24 + 25 = 0$
$9 + 24i - 16 - 18 - i24 + 25 = 0$

$-25 + 25 - 24i + 24i = 0$ $f(Z) = Z^2 - 6Z + 25$, $f(3 + i4) = 0$

The L.H.S. = R.H.S.

Solving the quadratic equation

$Z = \frac{6 \pm \sqrt{36 - 100}}{2} = \frac{6 \pm \sqrt{64}\sqrt{-1}}{2} = 3 \pm 4i$ ↗ $3 + 4i$ ↘ $3 - 4i$

but we know that if a polynomial equation with real coefficients has comple roots, they occur in conjugate pairs, therefore if $Z_1 = 3 + i4$, then the other root is the conjugate of Z_1, $Z_2 = 3 - i4$.
This statement is justified by solving the quadratic equation.

In a quadratic equation, the roots always appear in conjugate pairs if they are complex, therefore the other root would be the conjugate of $3 + i4$, i.e. $3 - i4$ and there is no need to solve the quadratic equation.

12. If $Z_1 = \overline{Z_2}$ then $a + b^2 - 3i = \overline{2 - ab^2 i}$, therefore $a + b^2 - 3i = 2 + ab^2 i$

Equating real and imaginary terms

$a + b^2 = 2$ and $ab^2 = -3$

$a - \frac{3}{a} = 2$, $b^2 = -\frac{3}{a}$

$a^2 - 2a - 3 = 0$ $a = \frac{2 \pm \sqrt{4 + 12}}{2} = 1 \pm 2$

$a = 3$ or $a = -1$ then

$b^2 = -\frac{3}{3}$ or $b^2 = \frac{-3}{-1} = 3$

$b^2 = -1$ or $b = \pm\sqrt{3}$

$b = \pm\sqrt{-1}$ or $b = \pm\sqrt{3}$

$b = \pm i$ but b is to be real, this value is not valid since it is complex.

Therefore

$a = -1 \qquad b = \pm\sqrt{3}$

The conjugate of Z is either denoted by $\overline{Z}$ or Z^* these read Z bar or Z star.

13. $\overline{Z} = Z^2$

Let $Z = x + iy$

$\overline{Z} = x - iy$ and $Z^2 = (x + iy)^2$

then $x - iy = (x + iy)^2 = x^2 + i^2y^2 + 2xyi$

$x - iy = x^2 - y^2 + 2xyi$

Equating real and imaginary terms $x = x^2 - y^2$ and $-y = 2xy$

$x = -\frac{1}{2}$ provided y is not zero

$-\frac{1}{2} = \frac{1}{4} - y^2$ and $y^2 = \frac{1}{4} + \frac{1}{2} = \frac{3}{4}$, $y = \pm\frac{\sqrt{3}}{2}$ $\qquad Z = -\frac{1}{2} \pm \frac{\sqrt{3}}{2}i$

Therefore $Z_1 = -\frac{1}{2} + \frac{\sqrt{3}}{2}i \qquad Z_2 = -\frac{1}{2} - \frac{\sqrt{3}}{2}i$

or $Z = -\frac{1}{2} \pm \frac{\sqrt{3}}{2}i$ the complex numbers which verify $\overline{Z} = Z^2$.

14. $Z_1 = x_1 + iy_1 \qquad Z_2 = x_2 + iy_2$

$$\begin{aligned} Z_1\overline{Z}_2 + \overline{Z}_1 Z_2 &= (x_1 + iy_1)(x_2 - iy_2) + (x_1 - iy_1)(x_2 + iy_2) \\ &= x_1x_2 + iy_1x_2 - ix_1y_2 + y_1y_2 + x_1x_2 - iy_1x_2 + iy_2x_1 + y_1y_2 \\ &= 2(x_1x_2 + y_1y_2) \end{aligned}$$

which is a real number.

15. (i) $$\begin{aligned} Z^2 + \overline{Z}^2 &= (x + iy)^2 + (x - iy)^2 \\ &= x^2 - y^2 + 2xyi + x^2 + y^2 - 2xyi \\ & 2x^2 \end{aligned}$$

(ii) $$\begin{aligned} \frac{Z + 1}{\overline{Z}} + \frac{\overline{Z} + 1}{Z} &= \frac{x + iy + 1}{x - iy} + \frac{x - iy + 1}{x + iy} \\ &= \frac{(x + iy + 1)(x + iy) + (x - iy + 1)(x - iy)}{x^2 + y^2} \\ &= \frac{x^2 + ixy + x + ixy - y^2 + iy + x^2 - ixy + x - ixy - y^2 - iy}{x^2 + y^2} \\ &= \frac{2x^2 - 2y^2 + 2x}{x^2 + y^2} = \frac{2(x^2 - y^2 + x)}{x^2 + y^2}. \end{aligned}$$

SOLUTIONS 7/8/9

1\. (i) $Z = 1$

$|Z| = 1$

$\arg Z = 0$

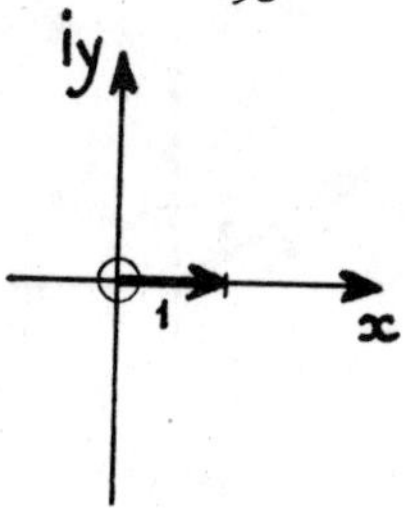

Fig.6 $z = 1\underline{/0^{o}}$

(ii) $Z = -1$

$|Z| = 1$

$\arg Z = \pi$

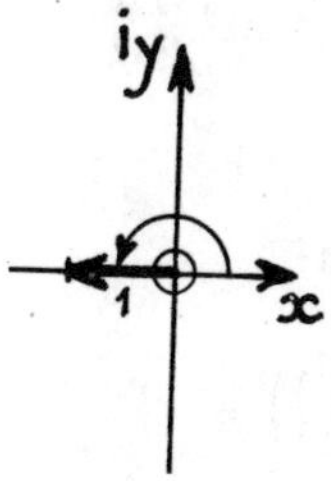

Fig.7 $z = 1\underline{/\pi}$

(iii) $Z = -i$

$|Z| = 1$

$\arg Z = -\pi/_2$

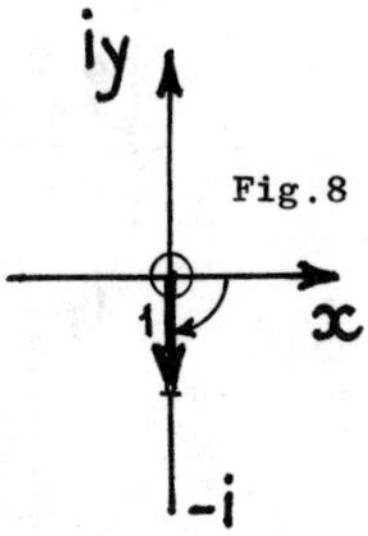

Fig.8 $z = 1\underline{/-\pi/_2}$

(iv) $Z = 1 + i\sqrt{3}$

$|Z| = \sqrt{1^2 + (\sqrt{3})^2}$

$= \sqrt{4}$

$= 2$

$\arg Z = \tan^{-1}\frac{\sqrt{3}}{1} = \pi/_3$

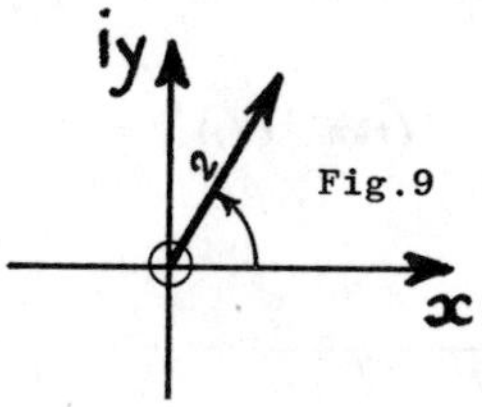

Fig.9 $z = 2\underline{/\pi/_3}$

(v) $Z = 1 - i\sqrt{3}$

$|Z| = \sqrt{1^2 + (-\sqrt{3})^2}$

$= \sqrt{4}$

$= 2$

$\arg Z = -\tan^{-1}\frac{\sqrt{3}}{1} = -\pi/_3$

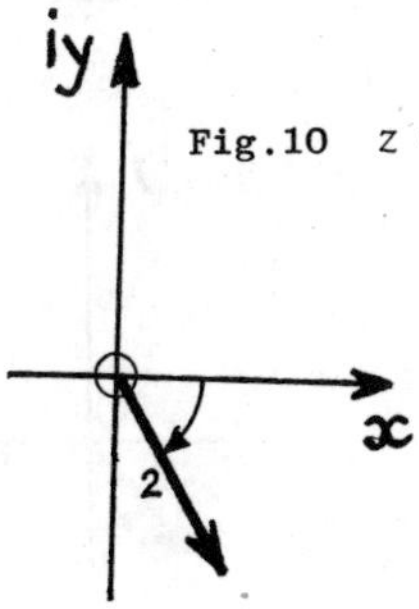

Fig.10 $z = 2\underline{/-\pi/_3}$

(vi) $Z = -1 + i\sqrt{3}$

$|Z| = \sqrt{(-1)^2 + (\sqrt{3})^2}$

$= \sqrt{4}$

$= 2$

$\arg Z = 2\pi/_3$

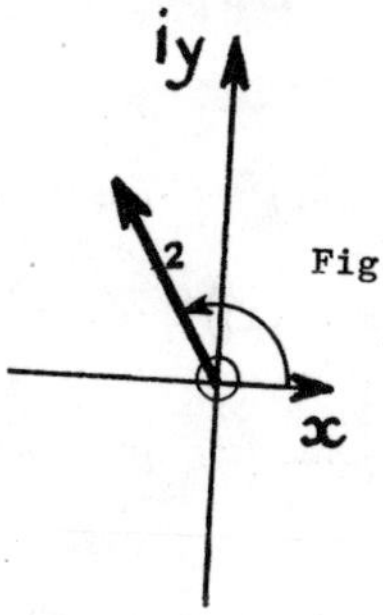

Fig.11 $z = 2\underline{/2\pi/_3}$

(vii) $Z = -1 - i\sqrt{3}$

$|Z| = \sqrt{(-1)^2 + (-\sqrt{3})^2}$

$= \sqrt{4}$

$= 2$

$\arg Z = 4\pi/_3$

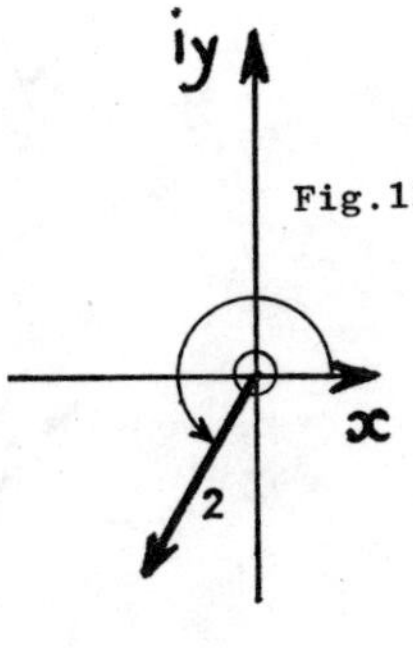

Fig.12 $z = 2\underline{/4\pi/_3}$

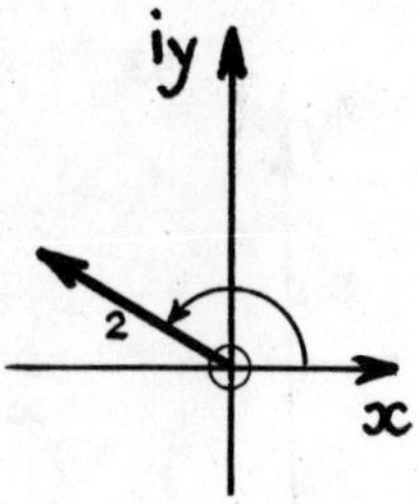

Fig.13 $z = 2\,\underline{/5\pi/6}$

(viii) $Z = -\sqrt{3} + i$

$|Z| = \sqrt{(-\sqrt{3})^2 + 1^2}$

$= \sqrt{4} = 2$

$\arg Z = \pi^c - \left(\tan^{-1}\frac{1}{\sqrt{3}}\right)^c = 5\pi^c/6$

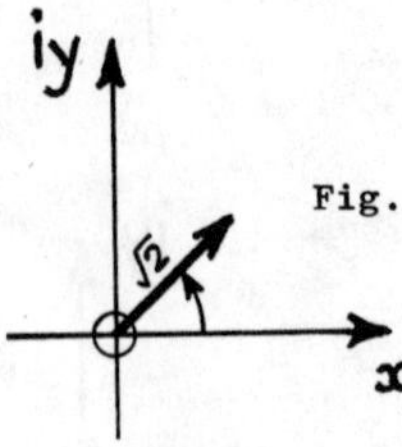

Fig.14 $z = \sqrt{2}\ \underline{/\pi/4}$

(ix) $Z = i + 1$

$|Z| = \sqrt{1^2 + 1^2}$

$= \sqrt{2}$

$\arg Z = \tan^{-1}\left(\frac{1}{1}\right) = \pi^c/4$

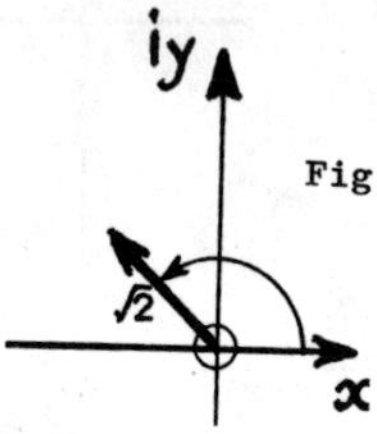

Fig.15 $z = \sqrt{2}\ \underline{/3\pi/4}$

(x) $Z = -1 + i$

$|Z| = \sqrt{(-1)^2 + 1}$

$= \sqrt{2}$

$\arg Z = \pi^c - \left(\tan^{-1}(1)\right)^c = \pi^c - \pi^c/4$

$= 3\pi^c/4$

(xi) $Z = -1 - i$

$|Z| = \sqrt{(-1)^2 + (-1)^2}$

$= \sqrt{2}$

$\arg Z = -(\pi - \tan^{-1} 1) = -3\pi^c/4$

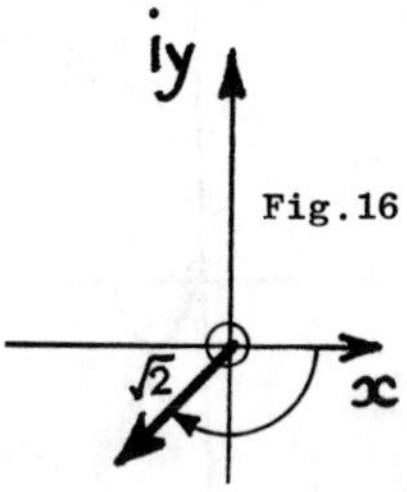

Fig.16 $z = \sqrt{2}\ \underline{/-3\pi/4}$

(xii) $Z = 1 - i$

$|Z| = \sqrt{1^2 + (-1)^2}$

$= \sqrt{2}$

$\arg Z = -\left(\tan^{-1}(1)\right)^c = -\pi^c/4$

Fig. 17

(xiii) $Z = i - 1$

$|Z| = \sqrt{(-1)^2 + 1^2}$

$= \sqrt{2}$

$\arg Z = \pi^c - \left(\tan^{-1}(1)\right)^c = \pi^c - \pi^c/4$

$= 3\pi^c/4$

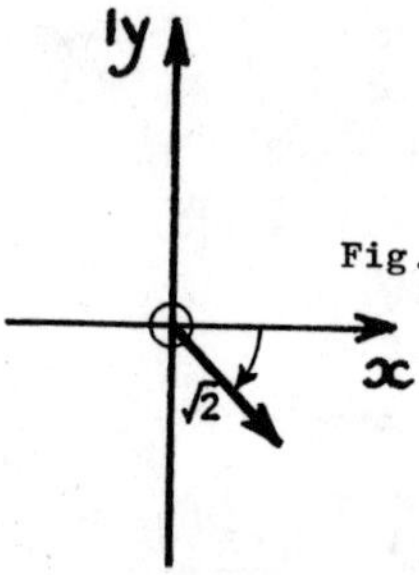

Fig.17 $z = \sqrt{2}\ \underline{/-\pi/4}$

Fig. 18

(xiv) $Z = \sqrt{3} - i$

$|Z| = \sqrt{(-1)^2 + (\sqrt{3})^2}$

$= \sqrt{4} = 2$

$\arg Z = -\tan^{-1}\left(\frac{1}{\sqrt{3}}\right) = -\pi^c/6$

Fig. 19

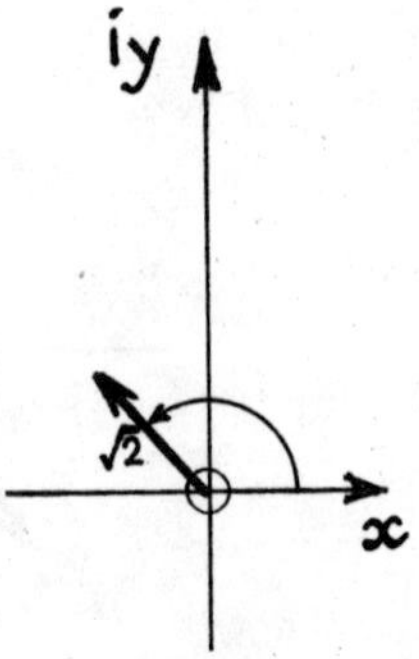

Fig.18 $z = \sqrt{2}\ \underline{/3\pi/4}$

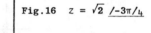

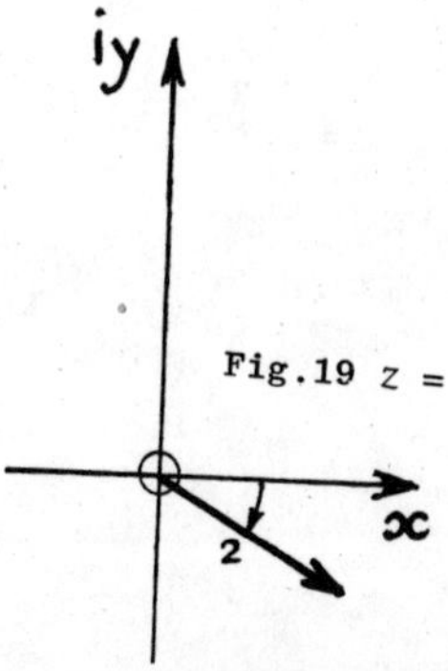

Fig.19 $z = 2\ \underline{/-\pi/6}$

(xv) $Z = -i - \sqrt{3}$

$|Z| = \sqrt{(-\sqrt{3})^2 \; (-1)^2}$

$= \sqrt{4} = 2$

$\arg Z = -\left(\pi^c - \tan^{-1}\frac{1}{\sqrt{3}}\right)$

$= -(\pi^c - \pi^c/6) = -5\pi^c/6$

Fig. 20

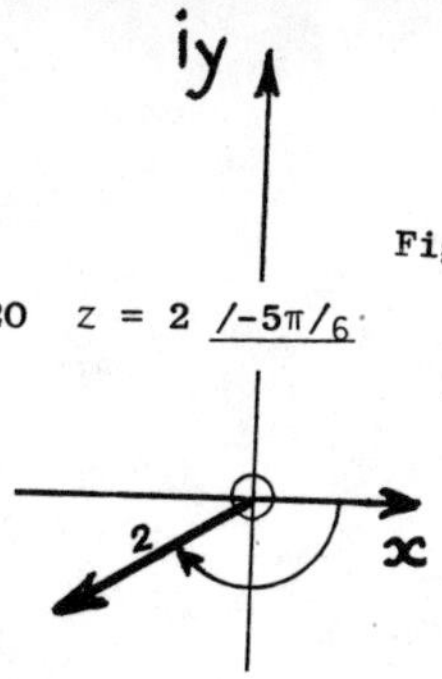

Fig.20 $z = 2 \; \underline{/-5\pi/6}$

(xvi) $Z = 2 + i3$

$|Z| = \sqrt{2^2 + 3^2}$

$= \sqrt{13}$

$\arg Z = \tan^{-1}(3/2) = 56^\circ \; 19'$

Fig. 21

Fig.21 $z = \sqrt{3} \; \underline{/56^\circ 19}$

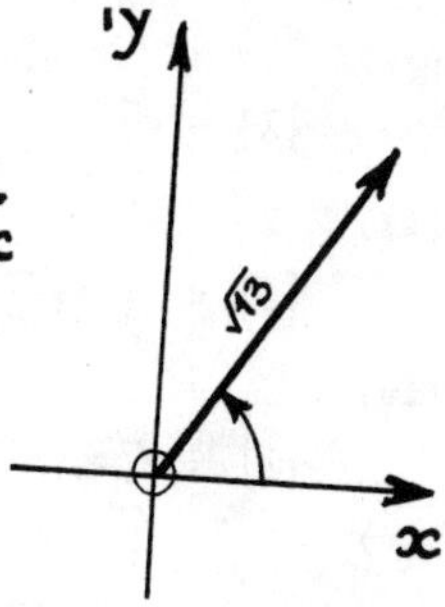

(xvii) $Z = -3 + i4$

$|Z| = \sqrt{(-3)^2 + (4)^2}$

$\arg Z = \pi^c - \tan^{-1}(4/3)$

Fig. 22

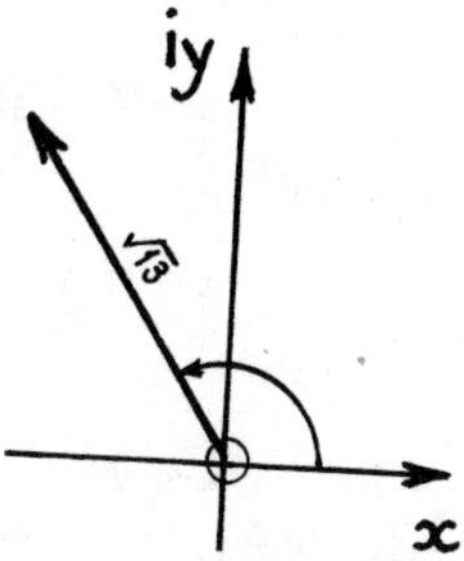

Fig.22 $z = \sqrt{13} \; \underline{/123^\circ} 41'$

(xviii) $Z = -2 - 4i$

$|Z| = \sqrt{(-2)^2 + (-4)^2}$

$= \sqrt{20}$

$\arg Z = -(\pi - \tan^{-1} 4/2) = -(\pi - 63^\circ 26')$

Fig. 23

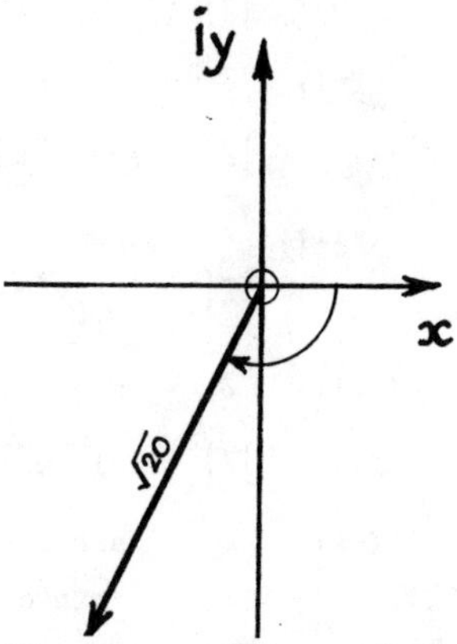

Fig.23 $z = \sqrt{20} \; \underline{/-116^\circ} 34'$

(xix) $Z = 3 - 2i$

$|Z| = \sqrt{3^2 + (-2)^2}$

$= \sqrt{13}$

$\arg Z = -\tan^{-1}(2/3) = -33^\circ 41'$

Fig. 24

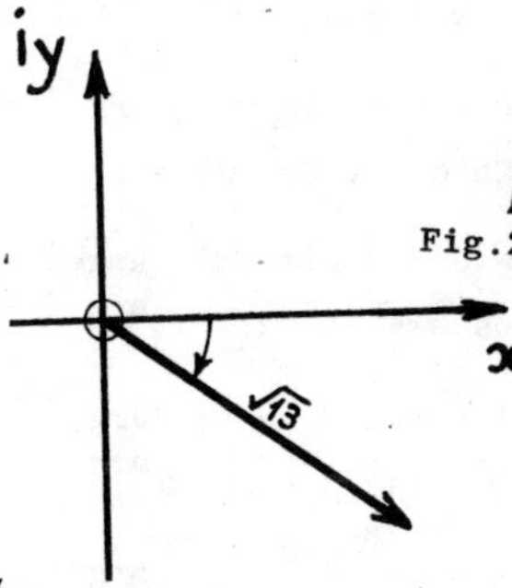

Fig.24 $z = \sqrt{13} \; \underline{/-33^\circ 41'}$

(xx) $Z = 5 - i3$

$|Z| = \sqrt{5^2 + (-3)^2}$

$= \sqrt{34}$

$\arg Z = -\tan^{-1}(3/5) = -30^\circ 58'$

Fig. 25

2. $(x + iy)^n = a + bi$

Let $W = a + bi$, $|W| = \sqrt{a^2 + b^2}$ $\arg W = \tan^{-1} b/a$ $\therefore W = \sqrt{a^2 + b^2}\,\underline{/\tan^{-1} b/a}$

Let $Z = x + iy$, $|Z| = \sqrt{x^2 + y^2}$ $\arg Z = \tan^{-1} y/x$. $\therefore Z = \sqrt{x^2 + y^2}\,\underline{/\tan^{-1} y/x}$

$\left[\sqrt{x^2 + y^2} \; \underline{/\tan^{-1} y/x}\right]^n = \sqrt{a^2 + b^2} \; \underline{/\tan^{-1}} \; b/a$

$W = (x^2 + y^2)^{n/2} \; \underline{/n \tan^{-1} y/x} = \sqrt{a^2 + b^2} \; \underline{/\tan^{-1}} \; b/a$

$\therefore \; a^2 + b^2 = (x^2 + y^2)^n$ and $n \tan^{-1} y/x = \tan^{-1} b/a = \arg W$

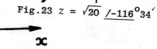

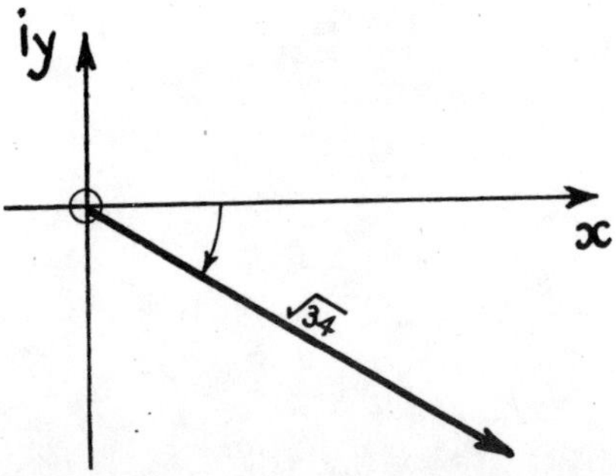

Fig.25 $z = \sqrt{34} \; \underline{/-30^\circ 58}$

3. (i) $Z = 3 + 4i$ $\quad$ $\arg Z = \tan^{-1}(^4/_3) = 53\overset{o}{\ }8$ $\quad$ Fig.26

$|Z| = \sqrt{3^2 + 4^2} = 5$

(ii) $Z = 3 - 4i$ $\quad$ $\arg Z = 2\pi^c - \tan^{-1}(^4/_3)$ $\quad$ Fig.27

$|Z| = \sqrt{3^2 + (-4)^2} = 5$

(iii) $Z = -3 + 4i$ $\quad$ $\arg Z = \pi^c - \tan^{-1}(^4/_3) = 116^o2$ $\quad$ Fig.28

$|Z| = \sqrt{(-3)^2 + (4)^2} = 5$

(iv) $Z = -3 - 4i$ $\quad$ $\arg Z = \pi^c + \tan^{-1}(^4/_3) = 233^o8$ $\quad$ Fig.29

$|Z| = \sqrt{(-3)^2 + (-4)^2} = 5$

(v) $Z = \sqrt{2} - i$ $\quad$ $\arg Z = 2\pi^c - \tan^{-1}\frac{1}{\sqrt{2}} = -35^o16$ $\quad$ Fig.30

$|Z| = (\sqrt{2})^2 + (-1)^2 = \sqrt{3}$

(vi) $Z = \sqrt{3} - i$ $\quad$ $\arg Z = 2\pi^c - \tan^{-1}\frac{1}{\sqrt{3}} = -30^o$ $\quad$ Fig.31

$|Z| = (\sqrt{3})^2 + (-1)^2 = 2$

(vii) $Z = \cos\alpha - i\sin\alpha$ $\quad$ $\arg Z = 2\pi^c - \tan^{-1}(\tan\alpha)$ $\quad$ Fig.32

$|Z| = \sqrt{\cos^2\alpha + (-\sin a)^2} = 1$ $\quad$ $= 2\pi^c - \alpha = -\alpha$

(viii) $Z = \sin\alpha + i\cos\alpha$ $\quad$ $\arg Z = \tan^{-1}\frac{\cos\alpha}{\sin\alpha}$ $\quad$ Fig.33

$|Z| = \sqrt{\sin^2\alpha + \cos^2\alpha} = 1$ $\quad$ $= \tan^{-1}(\cot\alpha)$

(ix) $Z = \sin\alpha - i\cos\alpha$ $\quad$ $\arg Z = 2\pi^c - \tan^{-1}(\cot\alpha)$ $\quad$ Fig.34

$|Z| = \sqrt{\sin^2\alpha + (-\cos^2\alpha)} = 1$

(x) $Z = \cos\alpha + i\sin\alpha$ $\quad$ $\arg Z = \tan^{-1}(\tan\alpha)$ $\quad$ Fig.35

$|Z| = \sqrt{\cos^2\alpha + \sin^2\alpha} = 1$ $\quad$ $\arg Z = \alpha$

(xi) $Z = 1 + i\tan\alpha$ $\quad$ $\arg Z = \tan^{-1}(\tan\alpha)$ $\quad$ Fig.36

$|Z| = \sqrt{1^2 + \tan^2\alpha} = \sec\alpha$ $\quad$ $\arg Z = \alpha$

(xii) $Z = 1 + i\cot\alpha$ $\quad$ $\arg Z = \tan^{-1}(\cot\alpha)$ $\quad$ Fig.37

$|Z| = \sqrt{1 + \cot^2\alpha} = \operatorname{cosec}\alpha$

(xiii) $Z = \tan\beta - i$ $\quad$ $\arg Z = 2\pi - \tan^{-1}\cot\beta$

$|Z| = \sqrt{\tan^2\beta + (-1)^2} = \sec\beta$ $\quad$ Fig.38

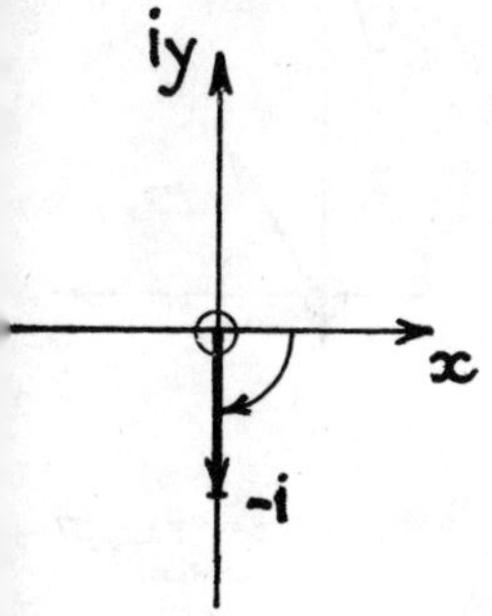

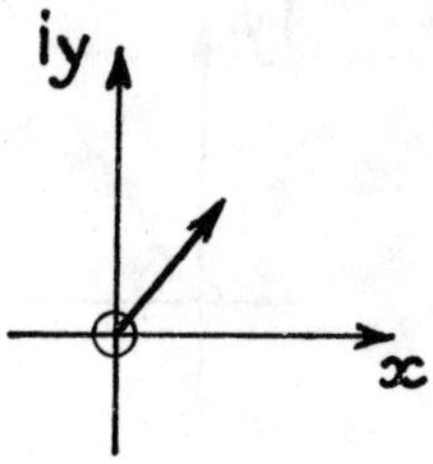

Fig.26

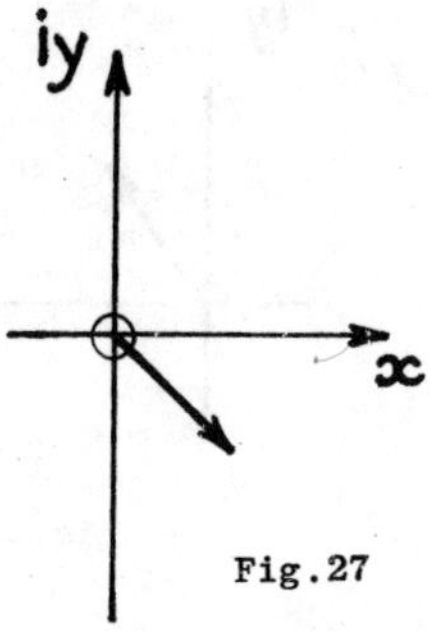

Fig.27

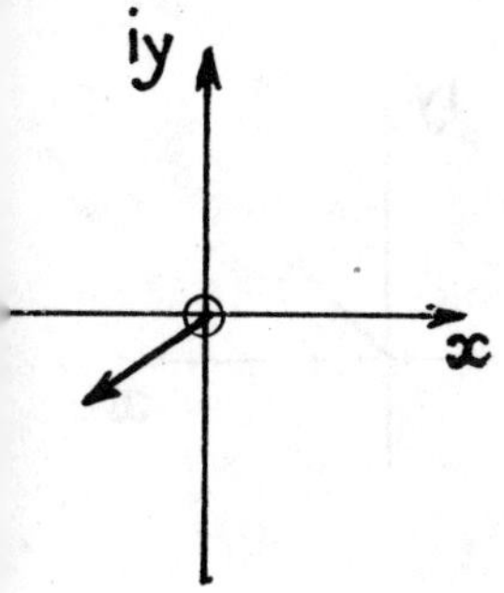

Fig.29

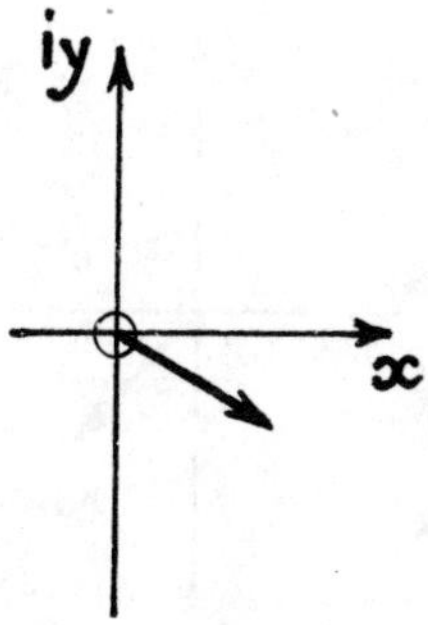

Fig.30

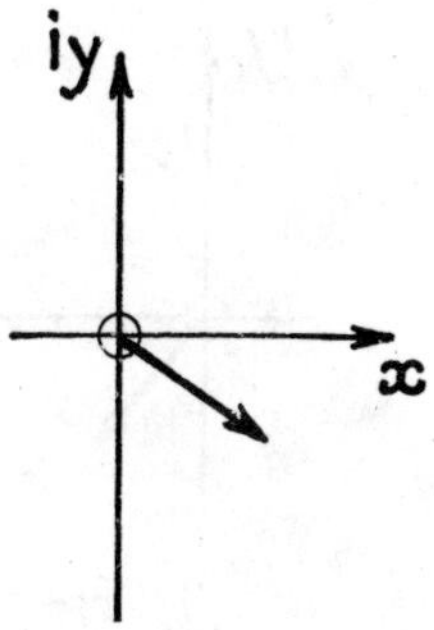

Fig.31

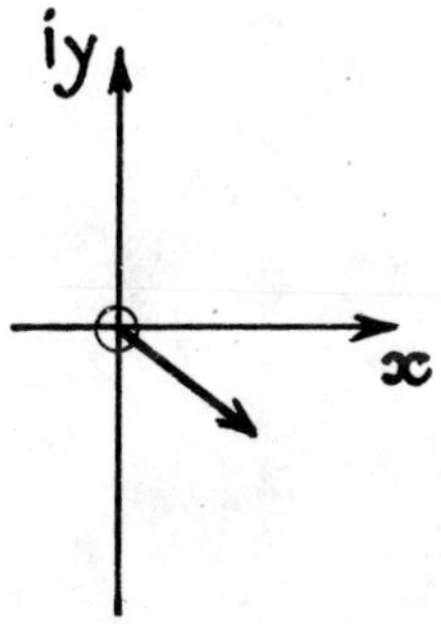

Fig.32

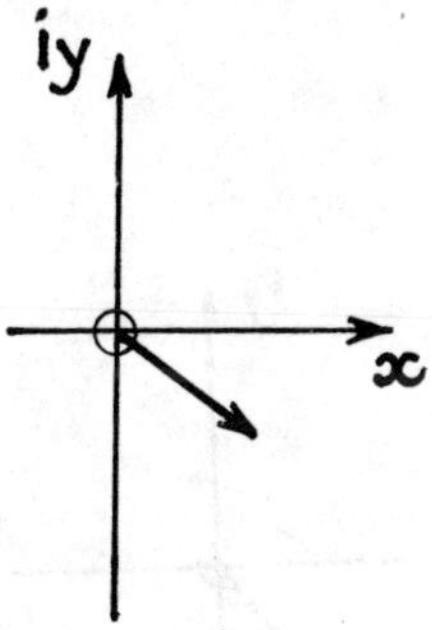

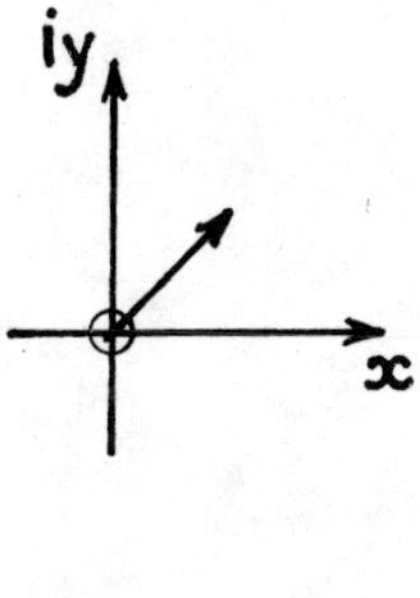

Fig.33

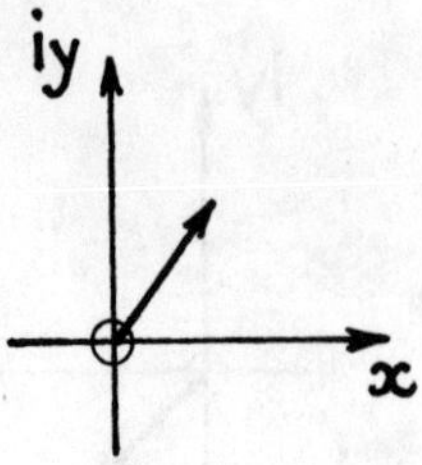

Fig.35

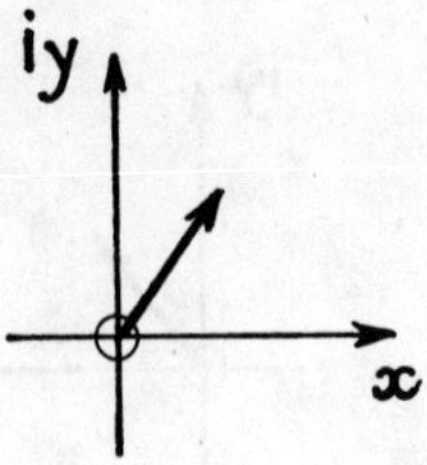

Fig.36

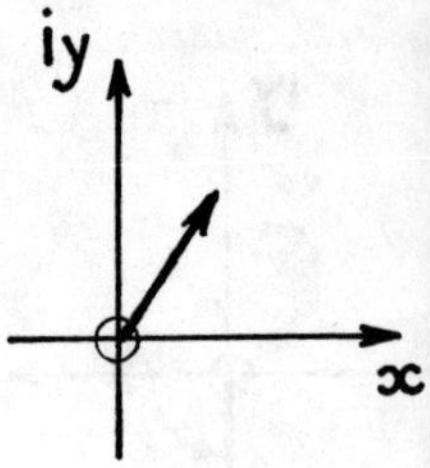

Fig.37

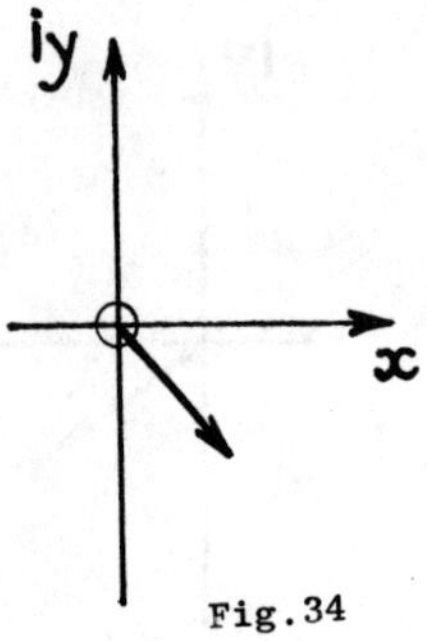

Fig.34

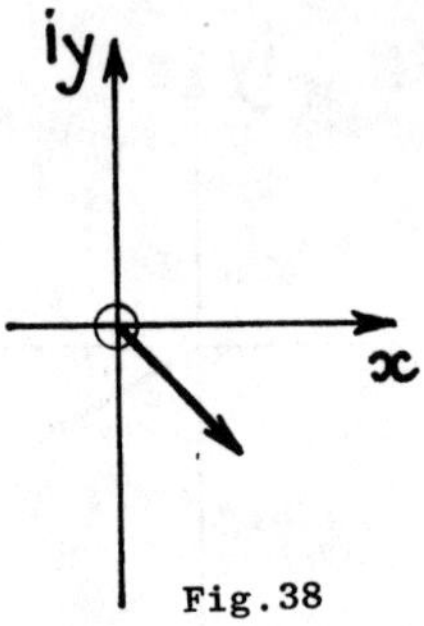

Fig.38

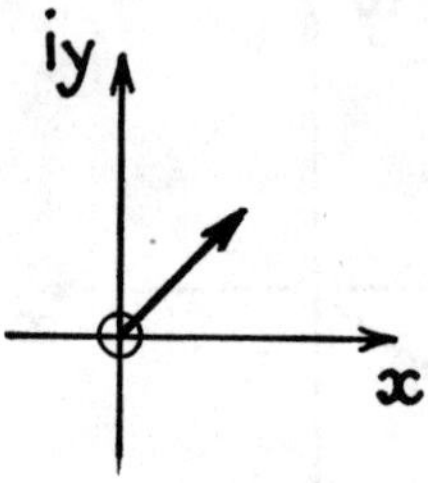

Fig.39

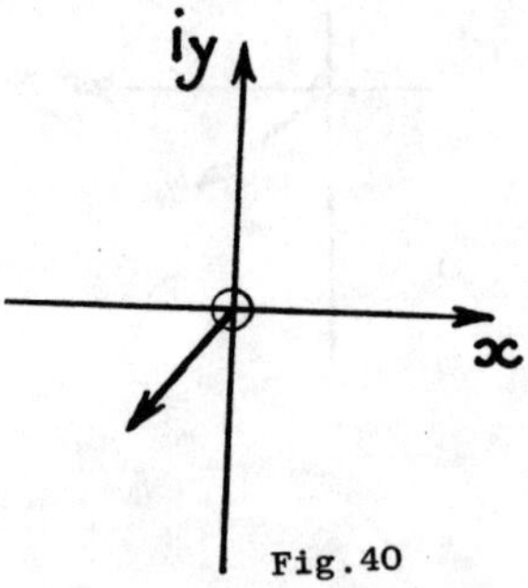

Fig.40

(xiv) $Z = \cos\alpha - i\sin\beta + i(\sin\alpha + i\cos\beta)$

$Z = (\cos\alpha - \cos\beta) - i(\sin\beta - \sin\alpha)$ If $\alpha > \beta$

$Z = (\cos\alpha - \cos\beta) + i(\sin\alpha - \sin\beta)$

$Z = 2\sin\frac{\alpha+\beta}{2}\sin\frac{\alpha-\beta}{2} + i\,2\cos\frac{\alpha+\beta}{2}\sin\frac{\alpha-\beta}{2}$

If $\alpha < \beta$

$$|Z| = \sqrt{4\sin^2\left(\frac{\alpha+\beta}{2}\right)\sin^2\left(\frac{\alpha-\beta}{2}\right) + 4\cos^2\left(\frac{\alpha+\beta}{2}\right)\sin^2\left(\frac{\alpha-\beta}{2}\right)}$$

$$= 2\sin\frac{\alpha-\beta}{2}\sqrt{\sin^2\left(\frac{\alpha+\beta}{2}\right) + \cos^2\left(\frac{\alpha+\beta}{2}\right)} = 2\sin\left(\frac{\alpha-\beta}{2}\right)$$

$$\arg Z = \tan^{-1}\frac{2\cos\left(\frac{\alpha+\beta}{2}\right)\sin\left(\frac{\alpha-\beta}{2}\right)}{2\sin\left(\frac{\alpha+\beta}{2}\right)\sin\left(\frac{\alpha-\beta}{2}\right)}$$

$\arg Z = \tan^{-1}\cot\left(\frac{\alpha+\beta}{2}\right)$ If $\alpha > \beta$

$\arg Z = \pi + \tan^{-1}\cot\left(\frac{\alpha+\beta}{2}\right)$ If $\alpha < \beta$

Fig. 40

(xv) $Z = 1 + r\cos\phi + ir\sin\phi$

$$|Z| = \sqrt{(1 + r\cos\phi)^2 + (r\sin\phi)^2}$$

$$= \sqrt{1 + 2r\cos\phi + r^2\cos^2\phi + r^2\sin^2\phi} = \sqrt{1 + 2r\cos\phi + r^2}$$

$$\arg Z = \tan^{-1}\left(\frac{r\sin\phi}{1 + r\cos\phi}\right)$$

Fig. 39

4 (i) $Z = 1\underline{/0^\circ}$, $Z = 1 + i0$

(ii) $Z = 3\underline{/-30^\circ}$, $Z = 3\cos 30^\circ - i3\sin 30^\circ = 3\frac{\sqrt{3}}{2} - i3 \times \frac{1}{2}$

(iii) $Z = 1\underline{/\pi/4}$, $Z = \cos \pi/4 + i \sin \pi/4 = \frac{1}{\sqrt{2}} + i\frac{1}{\sqrt{2}}$

(iv) $Z = 5\underline{/-\pi/2}$, $Z = 5\cos \pi/2 - i5 \sin \pi/2 = 0 - i5$

(v) $Z = 3\underline{/\pi}$, $Z = 3\cos \pi + i \sin \pi = -3 - i0$

(vi) $Z = \underline{/-180^\circ}$, $Z = \cos 180^\circ - i \sin 180^\circ = -1 + i0$

(vii) $Z = 3(\cos\theta + i\sin\theta) = 3\cos\theta + i3\sin\theta$

(viii) $Z = 1\underline{/353^\circ} = \cos 353^\circ + i \sin 353^\circ = 0.993 + i(-0.122)$

$Z = 0.993 - i0.122$

(ix) $Z = 3\underline{/360^\circ} = 3 + i0$

(x) $Z = 7\underline{/4\pi/3} = 7\cos 4\pi/3 + i7 \sin 4\pi/3$, $Z = -\frac{7}{2} - i7\frac{\sqrt{3}}{2}$

5 $Z = \frac{1 + 2i}{3 + 4i}$

$$Z = \frac{1+2i}{3+4i} \times \frac{3-4i}{3-4i} = \frac{3 + 6i - 4i - 8i^2}{3^2 - 4^2 i^2} = \frac{3 + 8 + 2i}{25} = \frac{11}{25} + \frac{2}{25}i$$

$$|Z| = \sqrt{\left(\frac{11}{25}\right)^2 + \left(\frac{2}{25}\right)^2} = \frac{1}{25}\sqrt{121 + 4} = \frac{1}{25}\sqrt{125} = \frac{1}{5}\sqrt{5} = 0.447$$

$$\arg Z = \tan^{-1}\frac{2/25}{11/25} = \tan^{-1} 2/11 = 10^\circ 18'$$

$$Z = 0.447\underline{/10^\circ 18'}$$

This problem can be solved alternatively

$$Z = \frac{1+2i}{3+4i}$$

$$|Z| = \frac{|1+2i|}{|3+4i|} = \frac{\sqrt{1^2 + 2^2}}{\sqrt{3^2 + 4^2}} = \frac{\sqrt{5}}{5}$$

$$\arg Z = \arg(1 + 2i) - \arg(3 + 4i)$$
$$= \tan^{-1} 2/1 - \tan^{-1} 4/3 = 63^\circ 26' - 53^\circ 8'$$
$$= 10^\circ 18'$$

6 $|Z| = \sqrt{2}$, $\arg Z = \pi/3$

(a) $Z = \sqrt{2}(\cos \pi/3 + i \sin \pi/3) = \sqrt{2}\left(\frac{1}{2} + i\frac{\sqrt{3}}{2}\right)$

$Z = \frac{\sqrt{2}}{2} + i\frac{\sqrt{6}}{2} = 0.707 + i1.23$

(b) $Z = \sqrt{2}\underline{/\pi/3}$ (c) $Z = \sqrt{2}e^{i\pi/3}$

7 $Z = 3 + i4$ $|Z| = \sqrt{3^2 + 4^2} = 5$ $\arg Z = \tan^{-1} 4/3 = 53^\circ 8'$

$$\frac{1}{Z} = \frac{1}{3 + i4}$$

$$\frac{1}{Z} = \frac{1}{3+i4} \times \frac{3-i4}{3-i4} = \frac{3 - i4}{3^2 - i^2 4^2} = \frac{3}{25} - i\frac{4}{25}$$

$\left|\frac{1}{Z}\right| = \sqrt{\frac{1}{25}}$ $\arg \frac{1}{Z} = -53^\circ 8'$

$Z^2 = (5\underline{/53^\circ 8'})^2 = 25\underline{/106^\circ}16'$, $Z^3 = (5\underline{/53^\circ 8'})^3 = 125\underline{/159^\circ}24'$

Z, $\frac{1}{Z}$, Z^2 and Z^3 are drawn in the diagrams but not to scale.

Fig. 41.

8. $Z_1 - Z_2 = (-1 - i) - (1 + i\sqrt{3})$
$= -1 - i - 1 - i\sqrt{3}$
$= -2 - i(1 + \sqrt{3}).$

$Z_1 + Z_2 = -1 - i + 1 + i\sqrt{3}$
$= -1 + \sqrt{3})i = (\sqrt{3} - 1)i.$

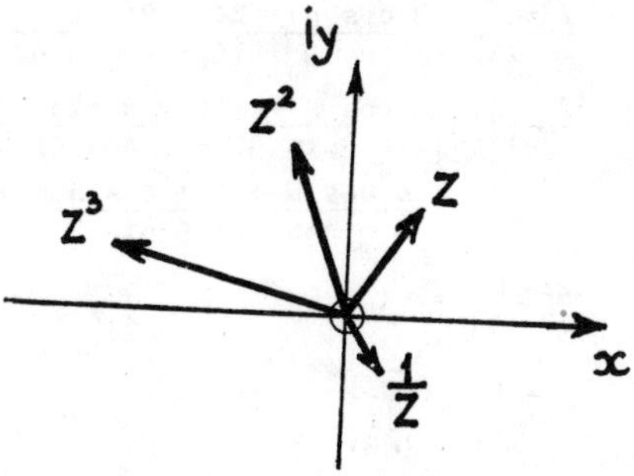

Fig.41 z, $\frac{1}{z}$, z^2 and z^3.

(i) $Z_1 = -1 - i$

$|Z_1| = \sqrt{(-1)^2 + (-1)^2} = \sqrt{2}$

$\arg Z_1 = \tan^{-1}\frac{1}{1} + \pi = \frac{\pi}{4} + \pi$

(ii) $Z_2 = 1 + i\sqrt{3}$

$|Z_2| = \sqrt{1 + (\sqrt{3})^2} = \sqrt{4} = 2$

$\arg Z_2 = \tan^{-1}\frac{\sqrt{3}}{1} = \pi/3$

(iii) $Z_1Z_2 = \sqrt{2}\,\underline{/5\pi/4} \quad 2\,\underline{/\pi/3} = 2\sqrt{2}\,\underline{/285^\circ}$

$|Z_1Z_2| = 2\sqrt{2}$

$\arg Z_1Z_2 = 285^\circ$

(iv) $\frac{Z_1}{Z_2} = \frac{\sqrt{2}\,\underline{/5\pi/4}}{2\,\underline{/\pi/3}} = \frac{\sqrt{2}}{2}\,\underline{/5\pi/4 - \pi/3} = \frac{1}{\sqrt{2}}\,\underline{/165^\circ}$

$\frac{Z_1}{Z_2} = \frac{1}{\sqrt{2}} \qquad \arg\frac{Z_1}{Z_2} = 165^\circ$

(v) $\frac{Z_2}{Z_1} = \frac{2\,\underline{/\pi/3}}{\sqrt{2}\,\underline{/5\pi/4}} = \frac{2}{\sqrt{2}}\,\underline{/-165^\circ}$

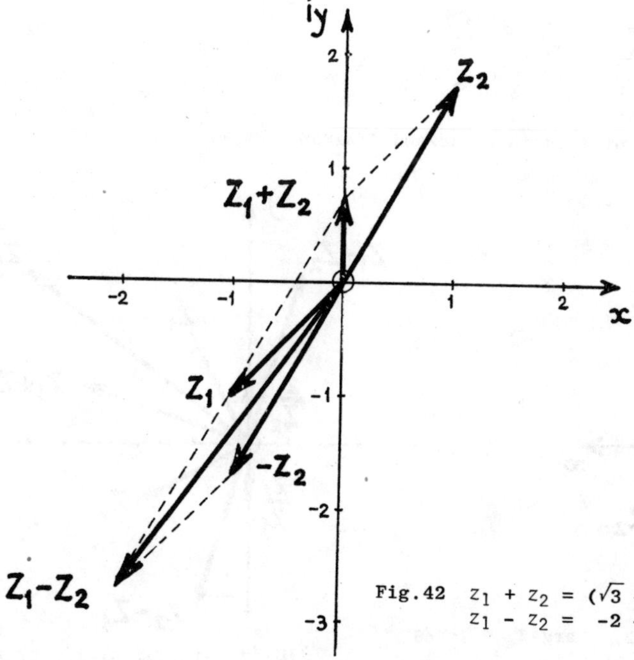

Fig.42 $z_1 + z_2 = (\sqrt{3} - 1)i$
$z_1 - z_2 = -2 - i(1 + \sqrt{3})$

9. If $Z = \cos\theta + i(1 + \sin\theta)$

$$\frac{2Z-i}{iZ-1} = \frac{2\cos\theta + 2i + 2i\sin\theta - i}{i\cos\theta + i^2 + i^2\sin\theta - 1} = \frac{2\cos\theta + i(1 + 2\sin\theta)}{(-2 - \sin\theta) + i\cos\theta}$$

$$\frac{|2Z-i|}{|iZ-1|} = \left|\frac{2\cos\theta + i(1 + 2\sin\theta}{(-2 - \sin\theta) + i\cos\theta)}\right| = \frac{\sqrt{4\cos^2\theta + (1 + 2\sin\theta)^2}}{\sqrt{(-2 - \sin\theta)^2 + \cos^2\theta}}$$

$$= \frac{4\cos^2\theta + 1 + 4\sin\theta + 4\sin^2\theta}{4 + \sin^2\theta + 4\sin\theta + \cos^2\theta} = \frac{4 + 1 + 4\sin\theta}{4 + 1 + 4\sin\theta} = 1$$

$$\frac{|2Z-i|}{|iZ-1|} = 1$$

10. $Z_1 = 1 + i\sqrt{3}$

$|Z_1| = \sqrt{1^2 + (\sqrt{3})^2} = 2$ $\qquad$ $\arg Z_1 = \tan^{-1}\frac{\sqrt{3}}{1} = \frac{\pi}{3}$

$Z_2 = \sqrt{3} - i$

$|Z_2| = \sqrt{(\sqrt{3})^2 + (-1)^2} = 2$ $\qquad$ $\arg Z_2 = -\tan^{-}\frac{1}{\sqrt{3}} = -\frac{\pi}{6}$

$Z_1Z_2 = 2\underline{/\pi/3}\quad 2\underline{/-\pi/6} = 4\underline{/\pi/6}$

$|Z_1Z_2| = 4$ $\qquad$ $\arg Z_1Z_2 = \frac{\pi}{6}$

$Z_1 + Z_2 = 1 + i\sqrt{3} + \sqrt{3} - i = (1 + \sqrt{3}) + i(\sqrt{3} - 1)$

$Z_1 - Z_2 = 1 + i\sqrt{3} - (\sqrt{3} - i) = (1 - \sqrt{3}) + i(\sqrt{3} + 1)$

$|Z_1 + Z_2| = \sqrt{(1 + \sqrt{3})^2 + (\sqrt{3} - 1)^2} = \sqrt{1 + 3 + 2\sqrt{3} + 3 + 1 - 2\sqrt{3}} = 2\sqrt{2}$

$\text{Arg}(Z_1 + Z_2) = +\tan^{-1}\frac{-1 + \sqrt{3}}{1 + \sqrt{3}} = 15^\circ$

$|Z_1 - Z_2| = \sqrt{(1 + \sqrt{3})^2 + (1 - \sqrt{3})^2} = \sqrt{1 + 3 + 2\sqrt{3} + 1 + 3 - 2\sqrt{3}} = 2\sqrt{2}$

$\text{Arg}(Z_1 - Z_2) = -\tan^{-1}\frac{1 + \sqrt{3}}{-1 + \sqrt{3}} = -75^\circ$

$$\frac{Z_2}{Z_1} = \frac{2\underline{/-\pi/6}}{2\underline{/\pi/3}} = 1\underline{/-\pi/2}$$

$$\frac{Z_1}{Z_2} = \frac{2\underline{/\pi/3}}{2\underline{/-\pi/6}} = 1\underline{/\pi/2}$$

Z_1, Z_2, Z_1Z_2, $Z_1 + Z_2$, $Z_1 - Z_2$

$\frac{Z_1}{Z_2}$, $\frac{Z_1}{Z_2}$ are plotted in an Argand diagram.

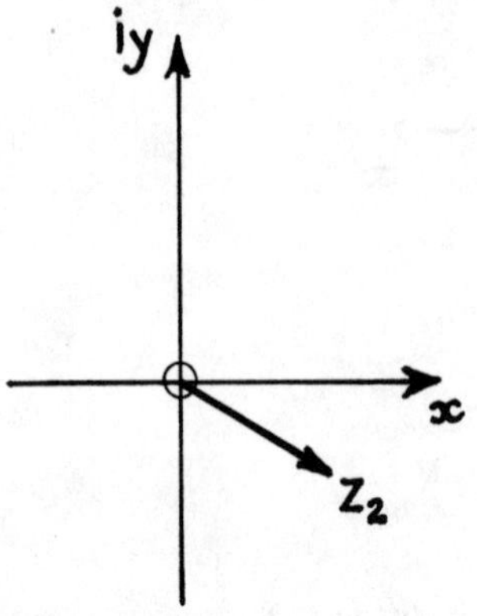

Fig.43 $|z_2| = 2$, $\arg z_2 = -\pi/6$

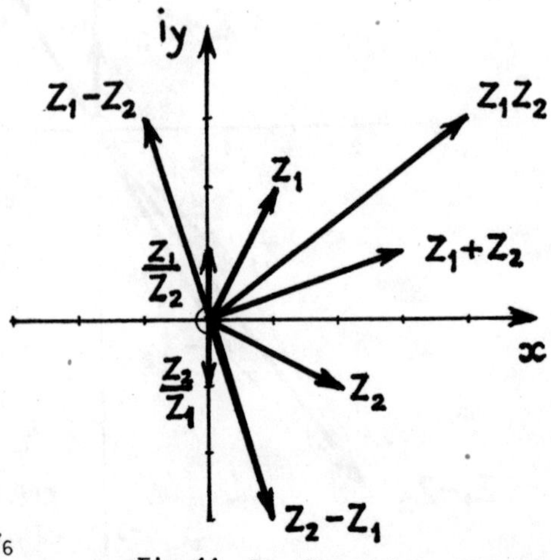

Fig.44 z_1, z_2, z_1z_2, $z_1 + z_2$, $z_1 - z_2$, $\frac{z_1}{z_2}$ and $\frac{z_2}{z_1}$

SOLUTIONS 10

1 (i) $Z_1 = 3e^{-i\pi/2} = 3\underline{/-\pi/2} = 3(\cos \pi/2 - i \sin \pi/2) = -3i$

(ii) $Z_2 = 5e^{i\pi} = 5\underline{/\pi} = 5(\cos \pi - i \sin \pi) = -5$

(iii) $Z_3 = e^{i\pi/4} = 1(\cos \pi/4 + i \sin \pi/4) = 1\underline{/\pi/4} = \frac{1}{\sqrt{2}} + i\frac{1}{\sqrt{2}}$

(iv) $Z_4 = e^{-i\pi/3} = 1(\cos \pi/3 - i \sin \pi/3) = 1\underline{/-\pi/3} = \frac{1}{2} - i\frac{\sqrt{3}}{2}$

(v) $Z_5 = e^{i3\pi/2} = 1\underline{/3\pi/2} = 1 \cos 3\pi/2 + i \sin 3\pi/2 = -i$

(vi) $Z_6 = 4e^{i5\pi/6} = 4\underline{/5\pi/6} = 4(\cos 5\pi/6 + i \sin 5\pi/6) = -2\sqrt{3} + i2$

(vii) $Z_7 = -3e^{-i11\pi/6} = -3\underline{/-11\pi/6} = -3\left(\cos 11\pi/6 - i \sin \frac{11\pi}{6}\right)$
$= \frac{3\sqrt{3}}{2} - i\frac{3}{2}$

(viii) $Z_8 = e^{-i\pi/4} = 1\underline{/-\pi/4} = (\cos \pi/4 - i \sin \pi/4) = \frac{1}{\sqrt{2}} - i\frac{1}{\sqrt{2}}$

(ix) $Z_9 = e^{-i3} = \cos 3^c - i \sin 3^c = \underline{/-3^c} = -0.99 - i0.14$
$Z_9 = -0.99 - 0.14$
$Z_9 = 1\underline{/-171^\circ 53'} = (\cos 171^\circ 53' - i \sin 171^\circ 53')$

(x) $Z_{10} = e^{-i} = \cos 1^c - i \sin 1^c = \underline{/-57^\circ 18'} = 0.54 - i0.84$

2 (i) $Z_1 = 0 - 3i$ $\quad |Z_1| = 3$
arg $Z_1 = -\pi/2$ $\quad Z_1 = 3e^{-i\pi/2}$

(ii) $Z_2 = -5 + 0i$ $\quad |Z_2| = 5$
arg $Z_2 = \pi$ $\quad Z_2 = 5^{i\pi}$

(iii) $Z_3 = \frac{1}{\sqrt{2}} + i\frac{1}{\sqrt{2}}$ $\quad |Z_3| = \sqrt{\left(\frac{1}{\sqrt{2}}\right)^2 + \left(\frac{1}{\sqrt{2}}\right)^2} = 1$
arg $Z_3 = \pi/4$ $\quad Z_3 = e^{i\pi/4}$

(iv) $Z_4 = \frac{1}{2} - i\frac{\sqrt{3}}{2}$ $\quad |Z_4| \sqrt{\left(\frac{\sqrt{3}}{2}\right)^2 + \left(\frac{1}{2}\right)^2}$
arg $Z_4 = -\pi/3$ $\quad Z_4 = e^{-i\pi/3}$

(v) $Z_5 = 0 - i$ $\quad |Z_5| = 1$
arg $Z_5 = -\pi/2$ $\quad Z_5 = e^{i3\pi/2}$

(vi) $Z_6 = -2\sqrt{3} + i2$ $\quad |Z_6| = \sqrt{(-2\sqrt{3})^2 + 2^2} = 4$
arg $Z_6 = 5\pi/6$ $\quad Z_6 = 4e^{i5\pi/6}$

(vii) $Z_7 = -\frac{3\sqrt{3}}{2} - i\frac{3}{2}$ $\quad |Z_7| = \sqrt{\left(-\frac{3\sqrt{3}}{2}\right)^2 + \left(-\frac{3}{2}\right)^2} = 3$
arg $Z_7 = 7\pi/6$ $\quad Z_7 = 3e^{i\pi/6}$

(viii) $Z_8 = \frac{1}{\sqrt{2}} - i\frac{1}{\sqrt{2}}$ $\quad |Z_8| = \sqrt{\left(\frac{1}{\sqrt{2}}\right)^2 + \left(\frac{-1}{\sqrt{2}}\right)^2} = 1$
arg $Z_8 = -\pi/4$ $\quad Z_8 = e^{-i\pi/4}$

(ix) $Z_9 = -0.99 - i0.14$ $\quad |Z_9| = \sqrt{(-0.99)^2 + (-0.14)^2} = 1$
arg $Z_9 = \pi + 0.1414$ $\quad Z_9 = e^{i3.3^c}$

(x) $Z_{10} = 0.54 - i0.84$ $\quad |Z_{10}| = \sqrt{(-0.54)^2 + (-0.84)^2} = 1$
arg $Z_{10} = -1^c$ $\quad Z_{10} = e^{-i}$

3. (i) $Z_1 = 3\underline{/-\pi/2} = 3e^{-i\pi/2}$

(ii) $Z_2 = 3\underline{/\pi} = 3e^{i\pi}$

(iii) $Z_3 = 1\underline{/-\pi/4} = e^{-i\pi/4}$

(iv) $Z_4 = 1\underline{/-\pi/3} = e^{-i\pi/3}$

(v) $Z_5 = 1\underline{/3\pi/2} = e^{i3\pi/2}$

(vi) $Z_6 = 4\underline{/5\pi/6} = 4e^{i5\pi/6}$

(vii) $Z_7 = -3\underline{/-11\pi/6} = -3e^{-i11/\pi}$

(viii) $Z_8 = 1\underline{/-\pi/4} = e^{-i\pi/4}$

(ix) $Z_9 = \underline{/171^\circ 53'} = e^{-i3}$

(x) $Z_{10} = \underline{/-57^\circ 18'} = e^{-i}$

When the complex number is expressed in the exponential form $re^{i\theta}$, θ is expressed only in radians, therefore, in question (ix) the argument $\theta = 171^\circ 53'$ must be converted into radians

$$\frac{171^\circ\ 53'\pi}{180^\circ} = 3^c \quad \text{where 'c' denotes radians.}$$

Similarly for $\theta = -57^\circ 18$ in (x) which is $\theta = \dfrac{-57^\circ 18' \times \pi}{180^\circ} = -1^c$.

4. To show that $\cos(\theta + \pi/4)e^{i\pi/4} - \sin(\theta - \pi/4)e^{-\pi/4} = (\cos\theta - \sin\theta)$

$\cos(\theta + \pi/4) = \cos\theta\cos\pi/4 - \sin\theta\sin\pi/4 = \frac{1}{\sqrt{2}}(\cos\theta - \sin\theta)$

$e^{i\pi/4} = \cos\pi/4 + i\sin\pi/4 = \frac{1}{\sqrt{2}}(1 + i)$

$\sin(\theta - \pi/4) = \sin\theta\cos\pi/4 - \sin\pi/4\cos\theta = \frac{1}{\sqrt{2}}(\sin\theta - \cos\theta)$

$e^{-i\pi/4} = \cos\pi/4 - i\sin\pi/4 = \frac{1}{\sqrt{2}}(1 - i)$

The left hand side

$$\frac{1}{\sqrt{2}}(1 + i)\frac{1}{\sqrt{2}}(\cos\theta - \sin\theta) - \frac{1}{\sqrt{2}}(\sin\theta - \cos\theta)\frac{1}{\sqrt{2}}(1 - i)$$

$$= \frac{1}{2}\{\cos\theta - \sin\theta + i\cos\theta - i\sin\theta - \sin\theta + \cos\theta + i\sin\theta - i\cos\theta\}$$

$$= \frac{2(\cos\theta - \sin\theta)}{2} = \cos\theta - \sin\theta.$$

Therefore $\cos(\theta + \pi/4)e^{i\pi/4} - \sin(\theta - \pi/4)e^{-i\pi/4} = \cos\theta - \sin\theta$.

5. $Z = \cos\pi/2 + i\sin\pi/2 = e^{i\pi/2}$

$Z^2 = \left(e^{i\pi/2}\right)^2 = e^{i\pi}$

$Z^3 = \left(e^{i\pi/2}\right)^3 = e^{i3\pi/2}$

6. $Z_1Z_2 = 3 + i4$ $\quad \frac{Z_1}{Z_2} = 5i \quad \left|\frac{Z_1}{Z_2}\right| = 5$

$|Z_1Z_2| = \sqrt{3^2 + 4^2} = 5$

$\arg Z_1Z_2 = \tan^{-1}\frac{4}{3} = 53^\circ 8'$

If $Z_1 = r_1 \angle\theta_1$ and $Z_2 = r_2 \angle\theta_2$ $\quad \arg\frac{Z_1}{Z_2} = 90^\circ$

$|Z_1Z_2| = |r_1 \angle\theta_1 \, r_2 \angle\theta_2| = r_1r_2 \quad \therefore \quad r_1r_2 = 5$

$\left|\frac{Z_1}{Z_2}\right| = \left|\frac{r_1\angle\theta_1}{r_2\angle\theta_2}\right| = \frac{r_1}{r_2} = 5 \quad \therefore \quad r_1 = 5r_2$ and $5r_2^{\,2} = 5$ $\boxed{r_2 = 1}$ and $\boxed{r_1 = 5}$

$\arg\frac{Z_1}{Z_2} = \arg(5i) = 90^\circ \qquad \arg Z_1Z_2 = 53^\circ 8' = \theta_1 + \theta_2$

$\theta_1 - \theta_2 = \arg Z_1 - \arg Z_2 = 90^\circ \quad \therefore \quad \theta_1 = 71^\circ 34'$ and

$\theta_2 = -18^\circ 26'$ $\quad Z_1 = 5\angle 71^\circ 34'$ and $Z_2 = 1\angle -18^\circ 26'$.

(i) $Z_1 = 5\cos 71^\circ 34' + i5\sin 71^\circ 34' = 1.581 + i4.743$
$Z_2 = \cos 18^\circ 26' - i\sin 18^\circ 26' = 0.949 - i0.316.$

(ii) $Z_1 = 5(\cos 71^\circ 34' + i\sin 71^\circ 34') = 5\angle 71^\circ 34'$ and $Z_2 = 1\angle -18^\circ 26'$.

(iii) $Z_1 = 5e^{i1.249}$ and $Z_2 = 1e^{-i0.322}$.

SOLUTIONS 11

1. It is required to verify that $1 - i$ is one of the square roots of $0 - 2i$.
Let $1 - i = \sqrt{0 - 2i} = \sqrt{-2i}$
Squaring up both sides
$$(1 - i)^2 = -2i$$
The left hand side may be expanded
$(1 - i)^2 = 1 + i^2 - 2i = 1 - 1 - 2i = -2i$
Therefore $1 - i$ is one of the square roots of $-2i$.
The other square root will be $-(1 - i) = -1 + i$ since $(1 - i)^2 = -2i$
gives $1 - i = \pm(-2i)^{\frac{1}{2}}$
$$\sqrt{-2i} = \pm(1 - i), \quad \sqrt{-2i} = 1 - i \quad \text{or} \quad \sqrt{-2i} = -1 + i.$$

2. Let $3 + i4 = \sqrt{-7 + 24i}$
Squaring up both sides $(3 + i4)^2 = -7 + 24i$.
The left hand side $(3 + i4)^2 = 9 + 24i + i^2 4^2 = 9 + 24i - 16$
$= -7 + 24i$
therefore $3 + i4$ is one of the square roots of $-7 + 24i$.
The other square root will be $-(3 + i4)$ or $-3 - i4$.

3. $(7 - i12)^2 = 49 - 168i - 144 = -95 - 168i$.
Therefore $7 - i12 = \sqrt{-95 - 168i}$.
The other square root is $-7 + i12$.

4. $(i)^2 = -1$. Therefore $i = \sqrt{-1}$ and the other square root is $-i$.

5. $(-3 - i4)^2 = -7 + 24i$.
$(-3 - i4)^2 = 9 + i^2 16 + 24i = -7 + 24i$
The other square root is $3 + i4$.

6. The square roots of the complex numbers can be found by using directly the formulae obtained in this section.

$$\sqrt{x + iy} = a + bi$$

then $a = \sqrt{\dfrac{\sqrt{x^2 + y^2} + x}{2}}$ $\quad b = \sqrt{\dfrac{\sqrt{x^2 + y^2} - x}{2}}$

but of course no one can remember these formulae, they have to be derived.

(i) $\sqrt{7 - i} = a + bi$

$$a = \sqrt{\frac{(7)^2 + (-1)^2 + 7}{2}} = 2.65, \quad b = \sqrt{\frac{(7)^2 + (-1)^2 - 7}{2}} = 0.189$$

therefore $7 - i = \pm(2.65 + 0.189i)$

(ii) $\sqrt{-1 + i} = a + bi$

$$a = \sqrt{\frac{\sqrt{(-1)^2 + (1)^2} + (-1)}{2}} = 0.455, \quad b = \sqrt{\frac{\sqrt{(-1)^2 + (1)^2} - (-1)}{2}} = 1.099$$

therefore $\sqrt{-1 + i} = \pm(0.455 + i1.099)$

(iii) $\sqrt{-3 - i4} = a + ib$

$$a = \sqrt{\frac{\sqrt{(-3)^2 + (-4)^2} - 3}{2}} = 1, \quad b = \sqrt{\frac{\sqrt{(-3)^2 + (-4)^2} - (-3)}{2}} = 2$$

therefore $\sqrt{-3 - i4} = \pm(1 + 2i)$

(iv) $\sqrt{4 - i3} = (a + ib)$

$$a = \sqrt{\frac{\sqrt{(4)^2 + (-3)^2} + 4}{2}} = \frac{3}{\sqrt{2}}, \quad b = \sqrt{\frac{\sqrt{(4)^2 + (-3)^2} - 4}{2}} = \frac{1}{\sqrt{2}}$$

therefore $\sqrt{4 - i3} = \pm\left(\frac{3}{\sqrt{2}} + \frac{1}{\sqrt{2}}i\right)$

(v) $\sqrt{-3 + 7i} = a + ib$

$$a = \sqrt{\frac{\sqrt{(-3)^2 + (7)^2} + (-3)}{2}} = 1.519, \quad b = \sqrt{\frac{\sqrt{(-3)^2 + (7)^2} - 3}{2}} = 2.304$$

therefore $\sqrt{-3 + 7i} = \pm(1.519 + 2.304i)$

(vi) $\sqrt{1 + 3i} = a + ib$

$$a = \sqrt{\frac{\sqrt{1^2 + 3^2} + 1}{2}} = 1.44, \quad b = \sqrt{\frac{\sqrt{1^2 + 3^2} - 1}{2}} = 1.04$$

therefore $\sqrt{1 + 3i} = \pm(1.44 + 1.04i)$

(vii) $\sqrt{4 + 7i} = a + ib$

$a = \sqrt{\frac{\sqrt{4^2 + 7^2} + 4}{2}} = 2.46$ $\qquad b = \sqrt{\frac{\sqrt{4^2 + 7^2} - 4}{2}} = 1.43$

therefore $\sqrt{4 + 7i} = \pm\ (2.46 + 1.43i)$

(viii) $\sqrt{-1 + 3i} = a + ib$

$a = \sqrt{\frac{\sqrt{(-1)^2 + 3^2} + (-1)}{2}} = 1.04,$ $\qquad b = \sqrt{\frac{\sqrt{(-1)^2 + 3^2} - (-1)}{2}} = 1.443$

therefore $\sqrt{-1 + 3i} = \pm(1.04 + i1.44)$

(ix) $\sqrt{-4 - i4} = a + ib$

$a = \sqrt{\frac{\sqrt{(-4)^2 + (-4)^2} + (-4)}{2}} = 0.91$ $\qquad b = \sqrt{\frac{\sqrt{(-4)^2 + (-4)^2} - (-4)}{2}} = 2.197$

therefore $\sqrt{-4 - i4} = \pm\ (0.91 + 2.197i)$

(x) $\sqrt{-6 + i} = a + ib$

$a = \sqrt{\frac{\sqrt{(-6)^2 + (1)^2} + (-6)}{2}} = 0.203$ $\qquad b = \sqrt{\frac{\sqrt{(-6)^2 + (1)^2} - (-6)}{2}} = 2.458$

therefore $\sqrt{-6 + i} = \pm(0.203 + 2.458i)$

All the above answers are plus or minus since when squaring up both sides it will give a positive answer.

7. Let $4 - i4 = \sqrt{-32i}$
Squaring up both sides $(4 - i4)^2 = -32i$
The L.H.S. $(4 - i4)^2 = 16 + i^2\ 16 - 32i = -32i$
therefore the square root of $-32i$ is $4 - i4$.
The other root will be $-4 + i4$.

8. $\sqrt{3 - 4} = \pm(a + ib)$

$a = \sqrt{\frac{\sqrt{3^2 + (-4)^2} + 3}{2}} = 2$ $\qquad b = \sqrt{\frac{\sqrt{3^2 + (-4)^2} - 3}{2}} = 1$

therefore $\sqrt{3 - i4} = \pm(2 + i).$

SOLUTIONS 12/13/14

1. (i) $\dfrac{\cos\phi - i\sin\phi}{\cos 2\phi + i\sin 2\phi} = \dfrac{(\cos\phi + i\sin\phi)^{-1}}{(\cos\phi + i\sin\phi)^{2}} = (\cos\phi + i\sin\phi)^{-3} = \cos 3\phi - i\sin 3\phi.$

(ii) $(\cos\theta - i\sin\theta)^7 = \cos 7\theta - i\sin 7\theta.$

(iii) $\dfrac{(\cos 2\theta - i\sin 2\theta)^4}{(\cos 3\theta + i\sin 3\theta)^3} = \dfrac{(\cos\theta + i\sin\theta)^{-8}}{(\cos\theta + i\sin\theta)^{9}} = (\cos\theta + i\sin\theta)^{-17} = \cos 17\theta - i\sin 17\theta.$

(iv) $(1 + \cos\theta + i\sin\theta)^3$ ....(1)

but $1 + \cos\theta = 1 + 2\cos^2\frac{\theta}{2} - 1 = 2\cos^2\frac{\theta}{2}$ ....(2)

and $\sin\theta = 2\sin\frac{\theta}{2}\cos\frac{\theta}{2}$ ....(3)

Substituting (2) and (3) in (1)

then $\left(2\cos^2\frac{\theta}{2} + i\,2\sin\frac{\theta}{2}\cos\frac{\theta}{2}\right)^3 = \left(2\cos\frac{\theta}{2}\right)^3\left(\cos\frac{\theta}{2} + i\sin\frac{\theta}{2}\right)^3$

$= 8\cos^3\frac{\theta}{2}\left(\cos\frac{3\theta}{2} + i\sin\frac{3\theta}{2}\right)$

$= 8\cos^3\frac{\theta}{2}\ \underline{/(3\theta/2)}$

2. (i) $Z + \frac{1}{Z} = \cos\theta + i\sin\theta + \cos\theta - i\sin\theta = 2\cos\theta$

$$\boxed{Z + \frac{1}{Z} = 2\cos\theta}$$

(ii) $Z - \frac{1}{Z} = \cos\theta + i\sin\theta - (\cos\theta - i\sin\theta) = 2i\sin\theta$

$$\boxed{Z - \frac{1}{Z} = 2i\sin\theta}$$

(iii) $Z^n + \frac{1}{Z^n} = (\cos\theta + i\sin\theta)^n + (\cos\theta + i\sin\theta)^{-n}$
$= \cos n\theta + i\sin n\theta + \cos n\theta - i\sin n\theta$

$$Z^n + \frac{1}{Z^n} = 2\cos n\theta$$

(iv) $Z^n - \frac{1}{Z^n} = (\cos\theta + i\sin\theta)^n - (\cos\theta + i\sin\theta)^{-n}$
$= \cos n\theta + i\sin n\theta - \cos n\theta + i\sin n\theta = 2i\sin n\theta$

$$\boxed{Z^n - \frac{1}{Z} = 2i\sin n\theta}$$

3. (i) $Z = (\cos 2\theta - i\sin 2\theta)^{\frac{1}{2}} = \cos\dfrac{2\theta + 2k\pi}{2} - i\sin\dfrac{2\theta + 2k\pi}{2}$

where $k = 0, 1$

when $k = 0$, $Z_1 = \cos\theta - i\sin\theta$

when $k = 1$, $Z_2 = \cos(\theta + \pi) - i\sin(\theta + \pi) = -\cos\theta + i\sin\theta$

therefore the roots of $\cos 2\theta - i\sin 2\theta$ are $\pm(\cos\theta - i\sin\theta)$.

(ii) $Z = (\cos 3\theta + i\sin 3\theta)^{\frac{1}{2}} = \underline{/(3\theta + 2k\pi)/2}$

where $k = 0, 1$

when $k = 0$ and when $k = 1$, then

$Z_1 = \underline{/3\theta/2}$ and $Z_2 = \underline{/3\theta/2 + \pi}$ are the two roots.

(iii) $Z = (\sin\theta + i\cos\theta)^{\frac{1}{2}} = \left(|\sin\theta + i\cos\theta|\ \underline{/\tan^{-1}\frac{\cos\theta}{\sin\theta}}\right)^{\frac{1}{2}}$

$= \left(\sqrt{\sin^2\theta + \cos^2\theta}\ \underline{/\tan^{-1}\cot\theta}\right)^{\frac{1}{2}}$

$Z = \left(\underline{/(\tan^{-1}\cot\theta + 2k\pi)}\right)^{\frac{1}{2}}$ $\quad k = 0, 1$

$Z_1 = \underline{/\tfrac{1}{2}\tan^{-1}\cot\theta}$ and $Z_2 = \underline{/\tfrac{1}{2}\tan^{-1}\cot\theta + \pi}$

$Z_1 = \underline{/\tfrac{1}{2}\tan^{-1}\tan\left(\frac{\pi}{2}-\theta\right)}$ $\qquad Z_2 = \underline{/\tfrac{1}{2}\tan^{-1}\tan\left(\frac{\pi}{2}-\theta\right) + \pi}$

$Z_1 = \underline{/\tfrac{1}{2}\left(\frac{\pi}{2}-\theta\right)}$ $\qquad Z_2 = \underline{/\tfrac{1}{2}\left(\frac{\pi}{2}-\theta\right) + \pi} = \underline{/\tfrac{1}{2}\left(\tfrac{5}{2}\pi-\theta\right)}$

(iv) $Z = (-i)^{\frac{1}{2}} = \left(\underline{/(3\pi/2 + 2k\pi)}\right)^{\frac{1}{2}}$

$Z_1 = \underline{/3\pi/4}$ and $Z_2 = \underline{/3\pi/4 + \pi} = \underline{/\tfrac{7}{4}\pi}$

(v) $Z = (i)^{\frac{1}{2}} = \left(\underline{/(\pi/2 + 2k\pi)}\right)^{\frac{1}{2}}$

$Z_1 = \underline{/\pi/4}$ and $Z_2 = \underline{/\pi/4 + \pi}$

$Z_1 = \underline{/\pi/4}$ and $Z_2 = \underline{/5\pi/4}$.

4 (i) $\sqrt{\dfrac{1+Z}{1-Z}} = \sqrt{\dfrac{1+\cos\theta + i\sin\theta}{1-\cos\theta - i\sin\theta}} = \sqrt{\dfrac{\sqrt{(1+\cos\theta)^2+\sin^2\theta}\ \underline{/\tan^{-1}\frac{\sin\theta}{1+\cos\theta}}}{\sqrt{(1-\cos\theta)^2+\sin^2\theta}\ \underline{/-\tan^{-1}\frac{\sin\theta}{1-\cos\theta}}}}$

$$= \left[\frac{(1+\cos\theta)^{\frac{1}{2}}}{(1-\cos\theta)^{\frac{1}{2}}}\ \frac{\underline{/\tan^{-1}\tan\theta/2}}{\underline{/-\tan^{-1}\cot\theta/2}}\right]^{\frac{1}{2}}$$

where $\dfrac{\sin\theta}{1+\cos\theta} = \dfrac{2\sin\theta/2\cos\theta/2}{1+2\cos^2\theta/2 - 1} = \dfrac{2\sin\theta/2\cos\theta/2}{2\cos^2\theta/2} = \tan\theta/2$

$\dfrac{\sin\theta}{1-\cos\theta} = \dfrac{2\sin\theta/2\cos\theta/2}{1-2\cos^2\theta/2+1} = \dfrac{2\sin\theta/2\cos\theta/2}{2(1-\cos^2\theta/2)} = \dfrac{2\sin\theta/2\cos\theta/2}{2\sin^2\theta/2}$

$= \cot\theta/2$

$$\sqrt{\frac{1+Z}{1-Z}} = \left(\left|\frac{\cos\theta/2}{\sin\theta/2}\right| \cdot \frac{\underline{/\tan^{-1}\tan\theta/2}}{\underline{/-\tan^{-1}\cot\theta/2}}\right)^{\frac{1}{2}}$$

$$= \sqrt{\cot\theta/2}\left(\frac{\underline{/\theta/2}}{\underline{/-(\pi/2-\theta/2)}}\right)^{\frac{1}{2}}$$

$$= \sqrt{\cot\theta/2}\left(\underline{/\theta/2 + \pi/2 - \theta/2}\right)^{\frac{1}{2}}$$

$$= \sqrt{\cot\theta/2}\ \underline{/\pi/4}$$

$$= \sqrt{\cot\theta/2}\ (\cos\pi/4 + i\sin\pi/4)$$

$$= \sqrt{\cot\theta/2}\left(\frac{1}{\sqrt{2}} + i\frac{1}{\sqrt{2}}\right)$$

$= \dfrac{1}{\sqrt{2}}\sqrt{\cot\theta/2}\,(1+i)$ where $0 < \theta < \pi$
$0 < \theta/2 < \pi/2$.

(ii) $\pi < \theta < 2\pi \Rightarrow \pi/2 < \theta/2 < \pi \Rightarrow -\pi/2 > -\theta/2 > -\pi \Rightarrow 0 > \pi/2 - \theta/2 > -\frac{\pi}{2}$
$\sin\theta$ is negative and $1-\cos\theta$ is always positive.

$$\left(\frac{1+Z}{1-Z}\right)^{\frac{1}{2}} = \sqrt{-\cot\theta/2}\left(\underline{/-\theta/2 - \pi/2 + \theta/2}\right)^{\frac{1}{2}}$$

$$= \sqrt{-\cot\theta/2}\ \underline{/-\pi/4} = \sqrt{-\cot\theta/2}\left(\frac{1}{\sqrt{2}} - i\frac{1}{\sqrt{2}}\right)$$

$$= \sqrt{\frac{-\cot\theta/2}{2}}\ (1-i).$$

5. $Z = \cos\theta + i \sin\theta$

$$Z^{-1} = \cos\theta - i\sin\theta$$

$$Z + \frac{1}{Z} = 2\cos\theta$$

$$\cos^3\theta = \frac{1}{2^3}\left(Z + \frac{1}{Z}\right)^3$$

$$= \frac{1}{8}\left(Z^3 + 3Z^2 . \frac{1}{Z} + 3Z . \frac{1}{Z^2} + \frac{1}{Z^3}\right)$$

$$= \frac{1}{8}\left(Z^3 + \frac{1}{Z^3} + 3\left(Z + \frac{1}{Z}\right)\right)$$

$$= \frac{1}{8}\,(2\cos 3\theta + 3(2\cos\theta)$$

$$\cos^3\theta = \frac{1}{4}\cos 3\theta + \frac{3}{4}\cos\theta$$

$$\boxed{\cos^3\theta = \frac{1}{4}(\cos 3\theta + 3\cos\theta)}$$

$$\cos^4\theta = \frac{1}{2^4}\left(Z + \frac{1}{Z}\right)^4$$

$$= \frac{1}{16}\left(Z^4 + 4Z^3\frac{1}{Z} + \frac{4 \times 3}{2} Z^2 \frac{1}{Z^2} + \frac{4 \times 3 \times 2}{1 \times 2 \times 3} Z \frac{1}{Z^3} + \frac{1}{Z^4}\right)$$

$$= \frac{1}{16} Z^4 + \frac{1}{4}\left(Z^2 + \frac{1}{Z^2}\right) + \frac{3}{8} + \frac{1}{16Z^4}$$

$$= \frac{1}{16}\left(Z^4 + \frac{1}{Z^4}\right) + \frac{1}{4}\left(Z^2 + \frac{1}{Z^2}\right) + \frac{3}{8}$$

$$= \frac{1}{16}\,2\cos 4\theta + \frac{1}{4}\,2\cos 2\theta + \frac{3}{8}$$

$$\boxed{\cos^4\theta = \frac{1}{8}\cos 4\theta + \frac{1}{2}\cos 2\theta + \frac{3}{8}}$$

$$Z^n = \cos n\theta + i\sin n\theta$$

$$Z^{-n} = \cos n\theta - i\sin n\theta$$

$$Z^n + Z^{-n} = 2\cos n\theta,\quad Z^n - Z^{-n} = 2i\sin n\theta$$

$$Z - \frac{1}{Z} = 2i\sin\theta$$

$$\sin^3\theta = \frac{1}{(2i)^3}\left(Z - \frac{1}{Z}\right)^3$$

$$= \frac{1}{8i^3}\left(Z^3 - 3Z^2 . \frac{1}{Z} + 3Z\frac{1}{Z^2} - \frac{1}{Z^3}\right)$$

$$= \frac{1}{8i}\left(Z^3 - \frac{1}{Z^3} - 3\left(Z - \frac{1}{Z}\right)\right)$$

$$= -\frac{1}{8i}\,(2i\,\sin 3\theta - 3\,(2i\sin\theta)$$

$$\sin^3\theta = -\frac{1}{4}\sin 3\theta + \frac{3}{4}\sin\theta$$

$$\boxed{\sin^3\theta = \frac{1}{4}(3\sin\theta - \sin 3\theta)}$$

$$\sin^4\theta = \frac{1}{(2i)^4}\left(Z - \frac{1}{Z}\right)^4$$

$$= \frac{1}{16i^4}\left(Z^4 - 4Z^2 + 6 - \frac{4}{Z^2} + \frac{1}{Z^4}\right)$$

$$\sin^4\theta = \frac{1}{16}\left(Z^4 + \frac{1}{Z^4}\right) - \frac{1}{4}\left(Z^2 + \frac{1}{Z^2}\right) + \frac{3}{8}$$

$$= \frac{1}{16}\,2\cos 4\theta - \frac{1}{4}\,2\cos 2\theta + \frac{3}{8}$$

$$\sin^4\theta = \frac{1}{8}\cos 4\theta - \frac{1}{2}\cos 2\theta + \frac{3}{8}$$

$$\boxed{\sin^4\theta = \frac{1}{8}\cos 4\theta - \frac{1}{2}\cos 2\theta + \frac{3}{8}}$$

$$\cos^5\theta = \frac{1}{2^5}\left(Z + \frac{1}{Z}\right)^5$$

$$= \frac{1}{32}\left(Z^5 + 5Z^4 . \frac{1}{Z} + \frac{5 \times 4}{2} Z^3 . \frac{1}{Z^2} + \frac{5 \times 4 \times 3}{1 \times 2 \times 3} Z^2 . \frac{1}{Z^3} + \frac{5 \times 4 \times 3 \times 2}{1 \times 2 \times 3 \times 4} . \frac{1}{Z^3} + \frac{1}{Z^5}\right.$$

$$= \frac{1}{32}\left(Z^5 + 5Z^3 + 10Z + 10\frac{1}{Z} + 5\frac{1}{Z^3} + \frac{1}{Z^5}\right)$$

$$= \frac{1}{32}\left(Z^5 + \frac{1}{Z^5}\right) + \frac{5}{32}\left(Z^3 + \frac{1}{Z^3}\right) + \frac{10}{32}\left(Z + \frac{1}{Z}\right)$$

$$= \frac{1}{32}\,2\cos 5\theta + \frac{5}{32}\,2\cos 3\theta + \frac{10}{32}\,2\cos\theta$$

$$\boxed{\cos^5\theta = \frac{1}{16}\cos 5\theta + \frac{5}{16}\cos 3\theta + \frac{5}{8}\cos\theta}$$

$$\sin^5\theta = \frac{1}{(2i)^5}\left(Z - \frac{1}{Z}\right)^5 = \frac{1}{32i^5}\left(Z^5 - 5Z^3 + 10Z - 10\frac{1}{Z} + 5\frac{1}{Z^3} - \frac{1}{Z^5}\right)$$

$$= \frac{1}{32i}\left(Z^5 - \frac{1}{Z^5}\right) - \frac{5}{32i}\left(Z^3 - \frac{1}{Z^3}\right) + \frac{1}{32i}\,10\left(Z - \frac{1}{Z}\right)$$

$$= \frac{2i}{32i}\sin 5\theta - \frac{5}{32i}\,2i\sin 3\theta + \frac{1}{32i} \times 10 \times 2i\sin\theta$$

$$\boxed{\sin^5\theta = \frac{1}{16}\sin 5\theta - \frac{5}{16}\sin 3\theta + \frac{5}{8}\sin\theta}$$

6\. (i) Let $Z^3 = \cos 3\theta - i \sin 3\theta$

$$Z = (\cos 3\theta - i \sin 3\theta)^{1/3} = \left(\underline{/- (3\theta + 2k\pi)}\right)^{1/3}$$

If $k = 0$	$k = 1$	$k = 2$
$Z_1 = \underline{/- \theta}$	$Z_2 = \underline{/- (\theta + 2\pi/3)}$	$Z_3 = \underline{/- (\theta + 4\pi/3)}$

The cube roots of Z^3 are

$$Z_1 = \cos \theta - i \sin \theta$$
$$Z_2 = \cos (\theta + 2\pi/3) - i \sin (\theta + 2\pi/3)$$
$$Z_3 = \cos (\theta + 4\pi/3) - i \sin (\theta + 4\pi/3).$$

(ii) Let $Z^3 = -i = \cos \pi/2 - i \sin \pi/2 = \cos 3\pi/2 + i \sin 3\pi/2 = \underline{/3\pi/2}$

$Z^3 = \underline{/3\pi/2}$ hence $Z = \left(\underline{/(3\pi/2 + 2k\pi)}\right)^{1/3}$

If $k = 0$	$k = 1$	$k = 2$
$Z_1 = \underline{/3\pi/6}$	$Z_2 = \underline{/3\pi/6 + 2\pi/3}$	$Z_3 = \underline{/3\pi/6 + 4\pi/3}$

The cube roots of $-i$ are:

$Z_1 = \underline{/\pi/2}$, $Z_2 = \underline{/7\pi/6}$ and $Z_3 = \underline{/11\pi/6}$.

(iii)

$$Z^3 = \sin \theta - i \cos \theta$$
$$Z^3 = \sqrt{\sin^2\theta + \cos^2\theta}\ \underline{/- \tan^{-1} \frac{\cos \theta}{\sin \theta}} = 1\underline{/- \tan^{-1} \tan\left(\frac{\pi}{2} - \theta\right)}$$
$$Z = \left(\underline{/- (\pi/2 - \theta)}\right)^{1/3} \qquad \cot \theta = \tan\left(\frac{\pi}{2} - \theta\right)$$
$$= \underline{/- \{(\pi/2 - \theta) + 2k\pi\}/3}$$

If $k = 0$	$k = 1$	$k = 2$
$Z_1 = \underline{/- (\pi/2 - \theta)/3}$	$Z_2 = \underline{/-(\pi/2 - \theta + 2\pi)/3}$	$Z_3 = \underline{/-(\pi/2 - \theta + 4\pi)/3}$

$$Z_1 = \cos\left(-\frac{1}{3}\theta + \frac{\pi}{6}\right) - i \sin\left(-\frac{1}{3}\theta + \frac{\pi}{6}\right)$$
$$Z_2 = \cos\left(-\frac{1}{3}\theta + \frac{5\pi}{6}\right) - i \sin\left(-\frac{1}{3}\theta + \frac{5\pi}{6}\right)$$
$$Z_3 = \cos\left(-\frac{1}{3}\theta + \frac{3\pi}{2}\right) - i \sin\left(-\frac{1}{3}\theta + \frac{3\pi}{2}\right).$$

7 (i) To find the roots of $Z^5 = 1$ .... (1)

Taking the fifth roots on both sides of (1)

$Z = 1^{1/5} = (\cos 0 + i \sin 0)^{1/5} = \underline{/2k\pi/5}$

$k = 0, 1, 2, 3, 4,$ or $k = 0, \pm 1, \pm 2.$

$Z_1 = 1\underline{/0^\circ}$	$Z_1 = 1\underline{/0^\circ}$
$Z_2 = 1\underline{/2\pi/5}$	$Z_2 = 1\underline{/2\pi/5}$
$Z_3 = 1\underline{/4\pi/5}$	$Z_3 = 1\underline{/-2\pi/5}$
$Z_4 = 1\underline{/6\pi/5}$	$Z_4 = 1\underline{/+4\pi/5}$
$Z_5 = 1\underline{/8\pi/5}$	$Z_5 = 1\underline{/-4\pi/5}$

The roots of $Z^5 = 1$

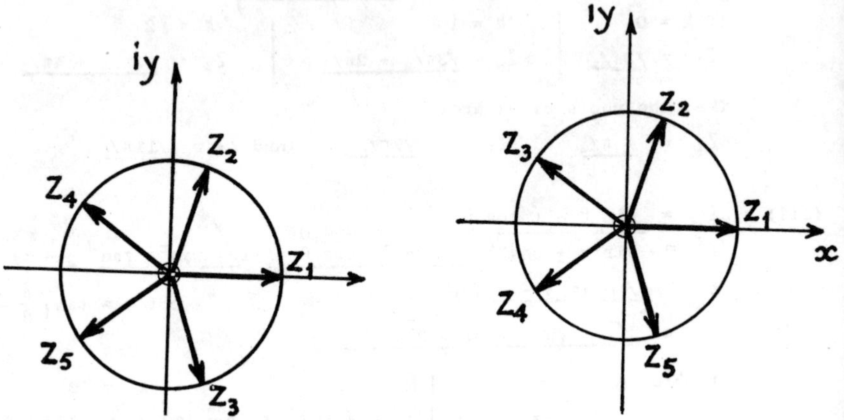

Fig.46 The roots of $z^5 = 1$, $k = 0, \pm 1, \pm 2$.

(ii) $Z^4 = 1 = \cos 0^\circ + i \sin 0^\circ = \underline{/0^\circ} = \underline{/0^\circ + 2k\pi} = \underline{/2k\pi}$

The four roots of unity $Z = 1^{1/4} = \underline{/(2k\pi)}^{1/4}$

$Z_1 = \underline{/0^\circ}$	$Z_2 = \underline{/2\pi/4}$	$Z_3 = \underline{/4\pi/4}$	$Z_4 = \underline{/6\pi/4}$
$Z_1 = 1\underline{/0^\circ}$	$Z_2 = 1\underline{/\pi/2}$	$Z_3 = 1\underline{/\pi}$	$Z_4 = \underline{/3\pi/2}$.

8. $(\cos\theta + i \sin\theta)^n + (\cos\theta + i\sin\theta)^{-n}$

$= \cos n\theta + i \sin n\theta + \cos n\theta - i \sin n\theta = 2 \cos n\theta$

9. $(\cos\theta + i \sin\theta)^3 = \cos 3\theta + i \sin 3\theta$

$= \cos^3\theta + 3\cos^2\theta\, i \sin\theta + 3\cos\theta\, i^2 \sin^2\theta + i^3 \sin^3\theta$

$= (\cos^3\theta - 3\cos\theta\sin^2\theta) + i(3\cos^2\theta\sin\theta - \sin^3\theta)$

Equating real and imaginary terms

$\cos 3\theta = \cos^3\theta - 3\cos\theta(1 - \cos^2\theta) = 4\cos^3\theta - 3\cos\theta$

$$\sin 3\theta = 3\cos^2\theta \sin\theta - \sin^3\theta = 3(1 - \sin^2\theta)\sin\theta - \sin^3\theta$$
$$= 3\sin\theta - 3\sin^3\theta - \sin^3\theta$$
$$\sin 3\theta = 3\sin\theta - 4\sin^3\theta$$

$$(\cos\theta + i\sin\theta)^4 = \cos 4\theta + i\sin 4\theta$$
$$= \cos^4\theta + 4\cos^3\theta\, i\sin\theta + \frac{4 \times 3}{1 \times 2}\cos^2\theta\, i^2 \sin^2\theta$$
$$+ \frac{4 \times 3 \times 2}{1 \times 2 \times 3}\cos\theta\, i^3 \sin^3\theta + i^4 \sin^4\theta$$
$$= \cos^4\theta + 4i\cos^3\theta\sin\theta - 6\cos^2\theta\sin^2\theta - 4i\sin^3\theta\cos\theta + \sin^4\theta$$

Equating real and imaginary terms

$$\cos 4\theta = \cos^4\theta - 6\cos^2\theta\,(1 - \cos^2\theta) + (1 - \cos^2\theta)^2$$
$$= \cos^4\theta - 6\cos^2\theta + 6\cos^4\theta + 1 - 2\cos^2\theta + \cos^4\theta$$
$$= 8\cos^4\theta - 8\cos^2\theta + 1$$
$$\sin 4\theta = 4\cos^3\theta\sin\theta - 4\sin^3\theta\cos\theta$$
$$= 4\sin\theta\cos\theta\,(\cos^2\theta - \sin^2\theta)$$

$$(\cos\theta + i\sin\theta)^5 = \cos 5\theta + i\sin 5\theta$$
$$= \cos^5\theta + 5\cos^4\theta\, i\sin\theta + \frac{5 \times 4}{1 \times 2}\cos^3\theta\, i^2\sin^2\theta$$
$$+ \frac{5 \times 4 \times 3}{1 \times 2 \times 3}\cos^2\theta\, i^3\sin^3\theta + \frac{5 \times 4 \times 3 \times 2}{1 \times 2 \times 3 \times 4}\cos\theta\, i^4\sin^4\theta$$
$$= \cos^5\theta + 5i\cos^4\theta\sin\theta - 10\cos^3\theta\sin^2\theta - 10i\cos^2\theta\sin^3\theta + 5\cos\theta\sin^4\theta$$

Equating real and imaginary terms

$$\cos 5\theta = \cos^5\theta - 10\cos^3\theta\sin^2\theta + 5\cos\theta\sin^4\theta$$
$$= \cos^5\theta - 10\cos^3\theta\,(1 - \cos^2\theta) + 5\cos\theta(1 - \cos^2\theta)^2$$
$$= \cos^5\theta - 10\cos^3\theta + 10\cos^5\theta + 5\cos\theta - 10\cos^3\theta + 5\cos^5\theta$$
$$= 16\cos^5\theta - 20\cos^3\theta + 5\cos\theta$$
$$\sin 5\theta = 5\cos^4\theta\sin\theta - 10\cos^2\theta\sin^3\theta.$$

10. $(\cos A + i\sin A)(\cos B + i\sin B)(\cos C + i\sin C)$

$$= \angle A \cdot \angle B \cdot \angle C = \angle A + B + C$$
$$= \angle \pi = -1$$

11. $Z^3 - 7 + 24i = 0$

$Z^3 = 7 - 24i$ or $Z = (7 - 24i)^{1/3} = 25^{1/3}\ \angle -(\tan^{-1}\frac{24}{7} + 2k\pi)/3$

when $k = 0$, $Z_1 = 25^{1/3}\ \angle -\frac{1}{3}\tan^{-1}\frac{24}{7} = 25^{1/3}\ \angle -24^\circ 35' = 2.924\ \angle -24^\circ 35'$

when $k = 1$. $Z_2 = 25^{1/3}\ \angle -(\tan^{-1}\frac{24}{7} + 2\pi)/3 = 2.924\ \angle -144^\circ 35'$

when $k = 2$, $Z_3 = 25^{1/3}\ \angle -(\tan^{-1}\frac{24}{7} + 4\pi)/3 = 2.924\ \angle -264^\circ 35'$.

$Z_1 = 2.66 - i\,1.22$, $Z_2 = -2.38 - i\,1.70$, and $Z_3 = -0.28 + i\,2.91$.

12. $Z^2 + 2i = 0$

$$Z^2 = -2i$$
$$Z = (-2i)^{\frac{1}{2}} = 2^{\frac{1}{2}}\ \angle (3\pi/2 + 2k\pi)/2$$

when $k = 0$, $Z_1 = \sqrt{2}\ \angle 3\pi/4$, and when $k = 1$ $Z_2 = \sqrt{2}\ \angle 7\pi/4$

$$Z_1 = \sqrt{2}\cos(3\pi/4) + i\sqrt{2}\sin(3\pi/4)$$
$$Z_1 = -1 + i$$
$$Z_2 = \sqrt{2}\cos(7\pi/4) + i\sqrt{2}\sin(7\pi/4)$$
$$Z_2 = 1 - i.$$

13. (i) $Z^3 - 1 = 0$

$Z^3 = 1$

$Z = 1^{1/3} = \left(\angle 0^\circ\right)^{1/3} = \angle 2k\pi/3$ where

$k = 0, 1, 2$

$Z_1 = \angle 0^\circ, \quad Z_2 = \angle 2\pi/3, \quad Z_3 = \angle 4\pi/3$

$Z_1 = 1$

$Z_2 = \cos 2\pi/3 + i \sin 2\pi/3 = -\frac{1}{2} + i\frac{\sqrt{3}}{2}$

$Z_3 = \cos 4\pi/3 + i \sin 4\pi/3 = -\frac{1}{2} - i\frac{\sqrt{3}}{2}$

$Z_1 + Z_2 + Z_3 = 1 - \frac{1}{2} + i\frac{3}{2} - \frac{1}{2} - i\frac{3}{2} = 0$

(ii) $Z^3 = i$

$Z = \left(\angle \pi/2\right)^{1/3}$

$Z = \left(\angle(\pi/2 + 2k\pi)\right)^{1/3}$ where $k = 0, 1, 2.$

$Z_1 = \angle \pi/6, \quad Z_2 = \angle 5\pi/6, \quad Z_3 = \angle 3\pi/2$

$Z_1 = \cos \pi/6 + i \sin \pi/6 = \frac{\sqrt{3}}{2} + i\frac{1}{2}$

$Z_2 = \cos 5\pi/6 + i \sin 5\pi/6 = -\frac{\sqrt{3}}{2} + i\frac{1}{2}$

$Z_3 = \cos 3\pi/2 + i \sin 3\pi/2 = -i$

Adding Z_1, Z_2, and Z_3

$Z_1 + Z_2 + Z_3 = \frac{\sqrt{3}}{2} + i\frac{1}{2} - \frac{\sqrt{3}}{2} + i\frac{1}{2} - i = 0$

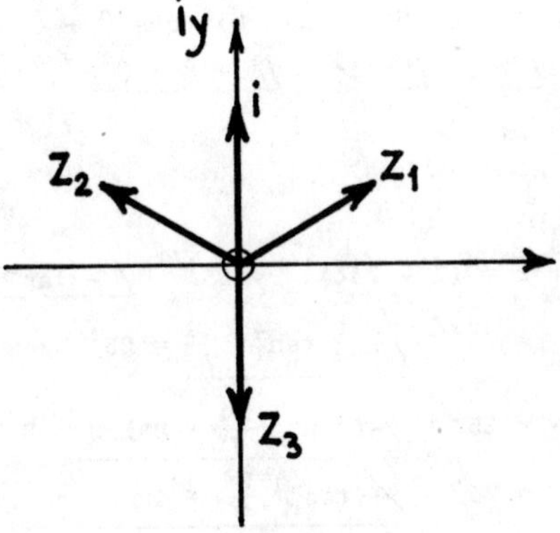

Fig.47 The roots of $z^3 - i = 0$.

(iii) $Z^3 + i = 0$

$Z^3 = -i$

$Z = \left(\underline{/3\pi/2}\right)^{1/3}$

$Z = \left(\underline{/(3\pi/2 + 2k\pi)}\right)^{1/3}$ $\quad k = 0, 1, 2$

$Z_1 = \underline{/\pi/2}$, $\quad Z_2 = \underline{/7\pi/6}$, $\quad Z_3 = \underline{/11\pi/6}$

$Z_1 = \cos \pi/2 + i \sin \pi/2 = i$

$Z_2 = \cos 7\pi/6 + i \sin 7\pi/6 = -\frac{\sqrt{3}}{2} - i\frac{1}{2}$

$Z_3 = - \cos 11\pi/6 + i \sin 11\pi/6 = +\frac{\sqrt{3}}{2} - i\frac{1}{2}$

Adding these vectors

$Z_1 + Z_2 + Z_3 = i - \frac{\sqrt{3}}{2} - i\frac{1}{2} + \frac{\sqrt{3}}{2} - i\frac{1}{2} = 0$

(iv) $Z^3 + 1 = 0$, $\quad Z^3 = -1$

$Z = (-1)^{1/3} = \left(\underline{/\pi + 2k\pi}\right)^{1/3} = \underline{/\frac{\pi + 2k\pi}{3}}$ where $k = 0, 1, 2.$

$Z_1 = \underline{/\pi/3}$, $\quad Z_2 = \underline{/\pi}$ $\quad Z_3 = \underline{/5\pi/3}$

$Z_1 = \cos \pi/3 + i \sin \pi/3$, $\quad Z_2 = \cos \pi + i \sin \pi$,
$Z_3 = \cos 5\pi/3 + i \sin 5\pi/3$

$Z_1 = \frac{1}{2} + i\frac{\sqrt{3}}{2}$, $\quad Z_2 = -1$, $\quad Z_3 = \frac{1}{2} - i\frac{\sqrt{3}}{2}$

Adding these vectors

$Z_1 + Z_2 + Z_3 = \frac{1}{2} + i\frac{\sqrt{3}}{2} - i + \frac{1}{2} - i\frac{\sqrt{3}}{2} = 0$

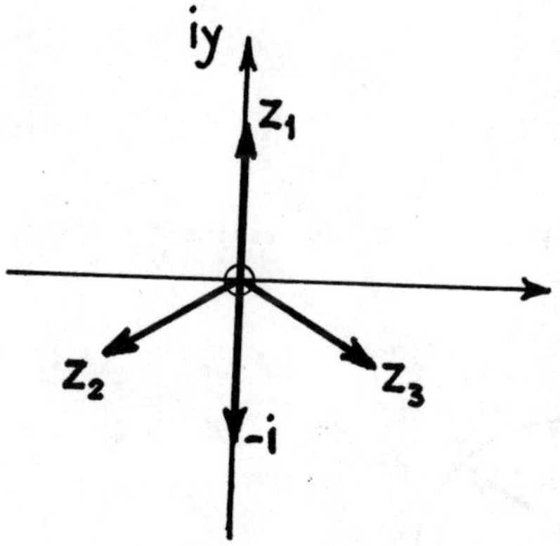

Fig.48 The roots of $z^3 + i = 0$.

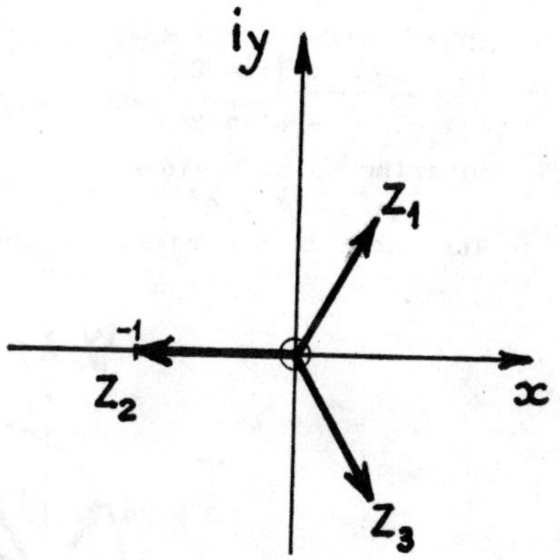

Fig.49 The roots of $z^3 + 1 = 0$.

SOLUTIONS 17/18

1. (i) $|Z| = 5$

$Z = x + iy$

$|Z| = \sqrt{x^2 + y^2} = 5$

Squaring up both sides $x^2 + y^2 = 5^2$

therefore, $C\ (0,0)$, $r = 5$ and the locus is a circle with centre the origin and radius 5.

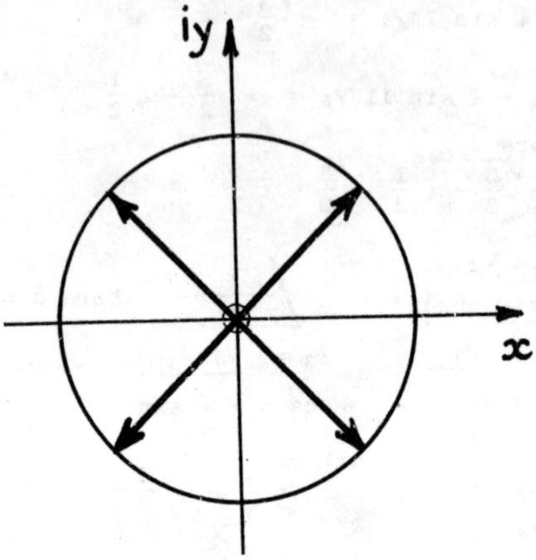

Fig.50 $|z| = 5$

(ii) $|Z - 1| = 2$

Substituting $Z = x + iy$

$|x + iy - 1| = 2$

$\sqrt{(x - 1)^2 + y^2} = 2$

Squaring up both sides

$(x - 1)^2 + y^2 = 2^2$

The locus is a circle with centre $C = (1, 0)$ and radius $r = 2$.

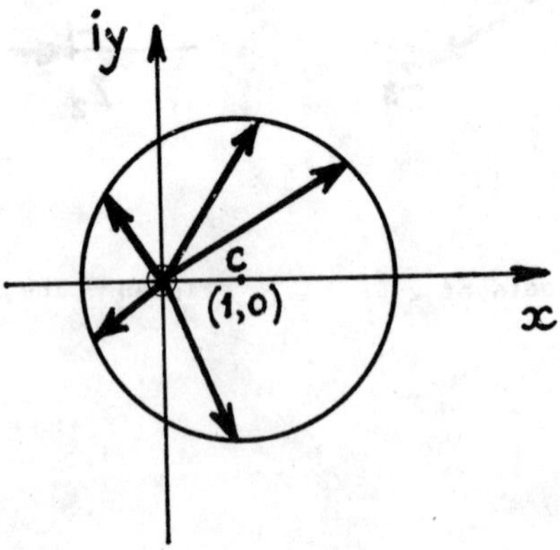

Fig.51 $|z - 1| = 2$

(iii) $|Z + 2| = 3$

Substituting $Z = x + iy$

$|x + iy + 2| = 3$

$\sqrt{(x + 2)^2 + y^2} = 3$

Squaring up both sides $(x + 2)^2 + y^2 = 3^2$

The locus is a circle with centre C (-2, 0) and radius $r = 3$.

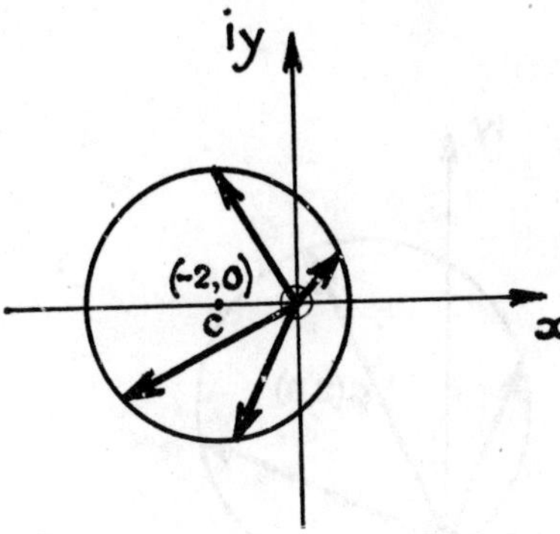

Fig.52 $|z + 2| = 3$

(iv) $|2Z - 1| = 3$

Substituting $Z = x + iy$

$|2(x + iy) - 1| = 3$

$\sqrt{(2x - 1)^2 + (2y)^2} = 3$

Squaring up both sides

$(2x - 1)^2 + 4y^2 = 3^2$

$\left(x - \frac{1}{2}\right)^2 + y^2 = \left(\frac{3}{2}\right)^2$

The locus is a circle with centre C $(\frac{1}{2}, 0)$ and radius $r = \frac{3}{2}$.

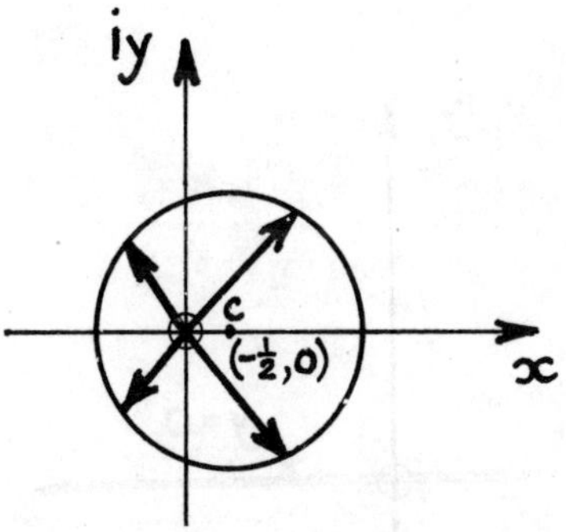

Fig.53 $|2z - 1| = 3$

(v) $|Z - 2 - 3i| = 4$

Substituting $Z = x + iy$

$|x + iy - 2 - 3i| = 4$

$|(x - 2) + i(y - 3)| = 4$

$\sqrt{(x - 2)^2 + (y - 3)^2} = 4$

Squaring up both sides $(x - 2)^2 + (y - 3)^2 = 4^2$

The locus is a circle with centre $C\ (2,3)$ and radius $r = 4$.

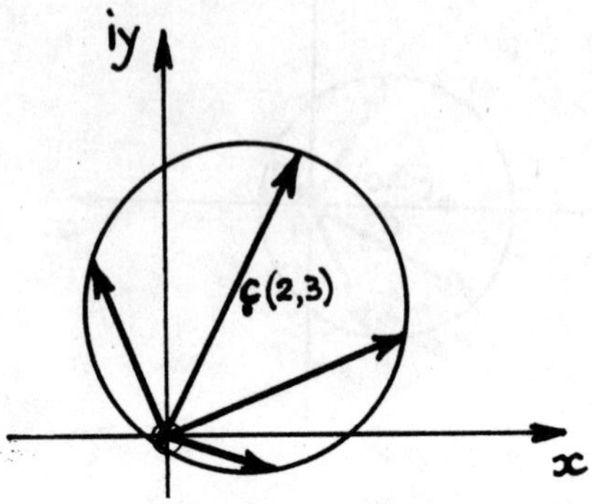

Fig.54 $|z - 2 - 3i| = 4$

(vi) $\arg Z = 0$

$\arg (x + iy) = \tan^{-1} y/x = 0$

$\frac{y}{x} = 0.$ Therefore, $y = 0$ is the locus which is a straight line.

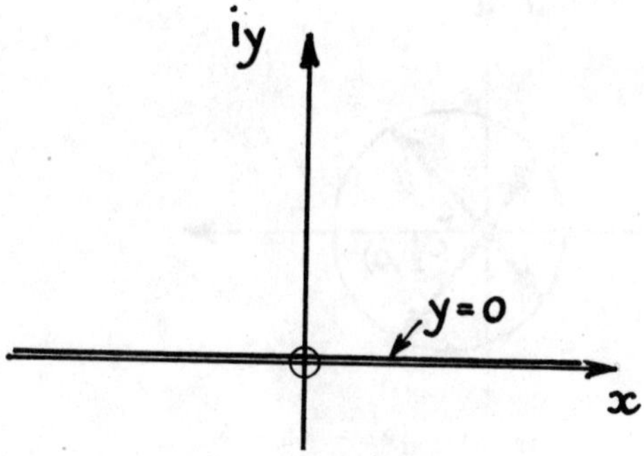

Fig.55 $\arg z = 0$

2. (i) $|z - 3| = |x + iy - 3| = \sqrt{(x-3)^2 + y^2}$

where $Z = x + iy$ if $|Z| \leq 1$

The greatest value = $AC = 4$
The least value = $BC = 2$

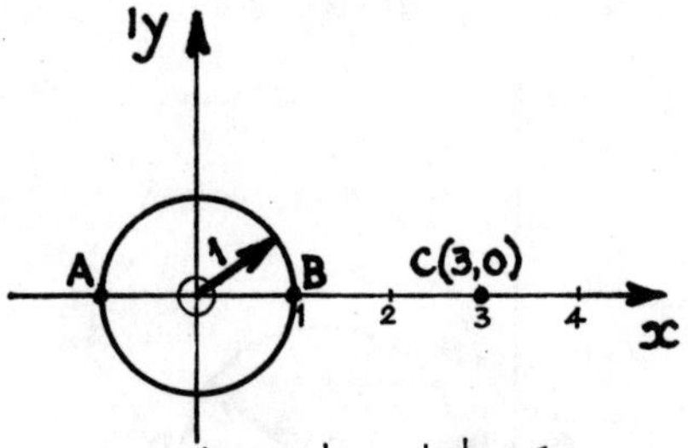

Fig.56 $|z - 3|$ if $|z| \leq 1$

(ii) $|z + 2| = |x + iy + 2| = \sqrt{(x+2)^2 + y^2}$

where $z = x + iy$ if $|Z| \leq 1$

The greatest value = 3 = CB
The least value = 1 = AC

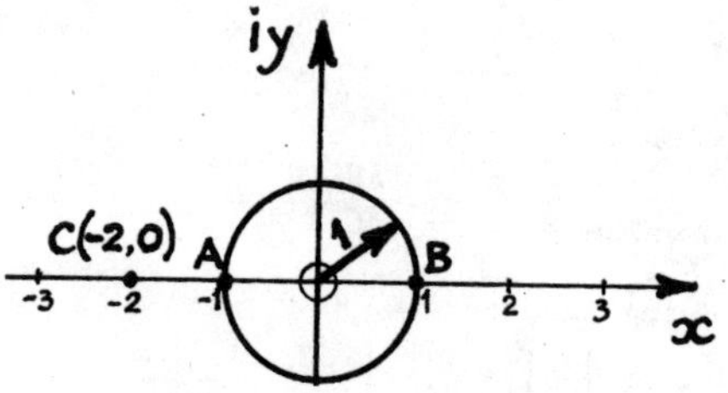

Fig.57 $|z + 2|$ if $|z| \leq 1$

(iii) $|x + iy| = \sqrt{x^2 + y^2}$ where $z = x + iy$

$|z - 5| \leq 2$

$|x + iy - 5| \leq 2$

$(x-5)^2 + y^2 \leq 2^2$

The greatest value = $AB = 7$
The least value = $AD = 3$

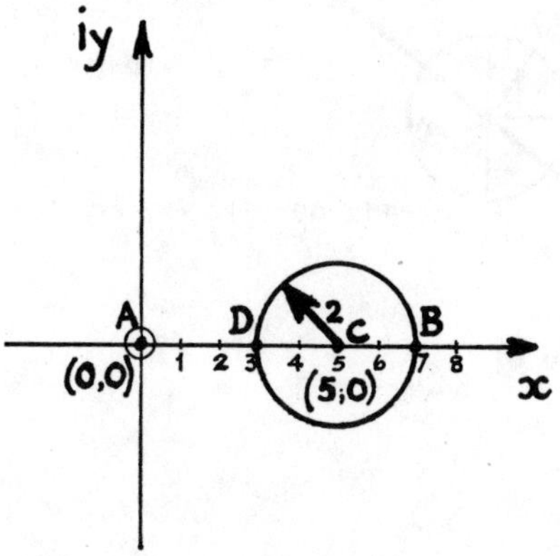

(iv) $|z + 1| = |x + iy + 1| = \sqrt{(x + 1)^2 + y^2}$

$|z - 4| \leqslant 3$

$(x + iy - 4) \leqslant 3$

$(x - 4)^2 + y^2 \leqslant 3^2$

The greatest value = 8 = AD

The least value = 2 = AB

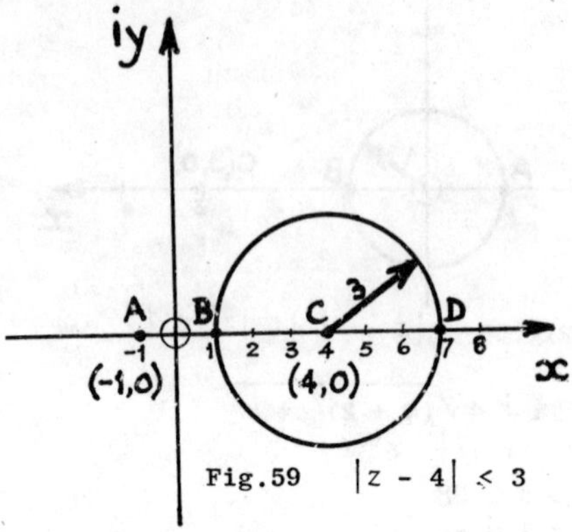

Fig.59 $|z - 4| \leqslant 3$

(v) $|z - 4| = \sqrt{(x - 4)^2 + y^2}$

$|z + 3i| \leqslant 1$

$(x + iy + 3i) \leqslant 1$

$x^2 + (y + 3)^2 \leqslant 1$

The greatest value = 6 = AB

The least value = 4 = AC

There are the following theorems

$|Z_1 + Z_2| \leqslant |Z_1| + |Z_2|$

$||Z_1| - |Z_2|| \leqslant |Z_1 + Z_2|$

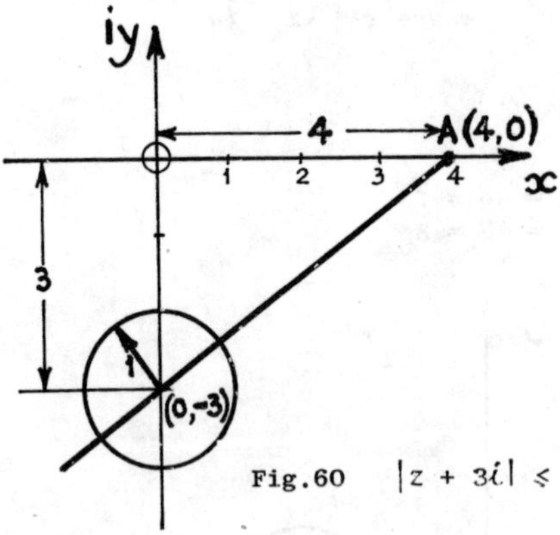

Fig.60 $|z + 3i| \leqslant 1$

3 (i) $z^3 = i$. Cube rooting both sides, we have

$z = i^{1/3}(\underline{/(\pi/2 + 2k\pi)})^{1/3} \quad k = 0.\ 1.\ 2$

by applying De Moivre's theorem for rational value of n

then the roots $z_1 = \underline{/\pi/6}$ $z_2 = \underline{/5\pi/6}$, $z_3 = \underline{/11\pi/6}$.

These are shown in the Argand diagram.

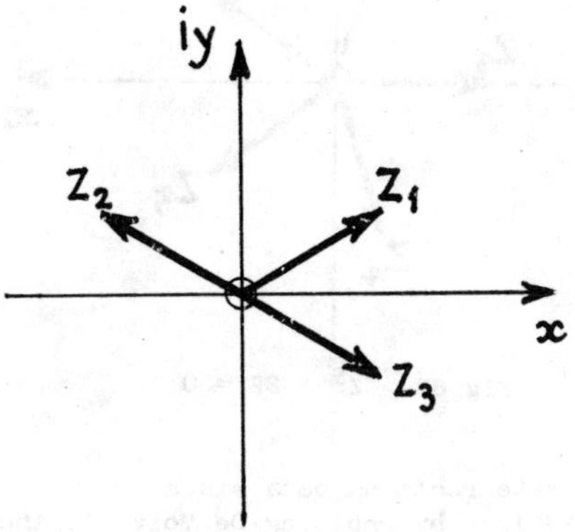

Fig.61 $z_1 = \underline{/\pi/6}$, $z_2 = \underline{/5\pi/6}$, $z_3 = \underline{/11\pi/6}$

(ii) $z^4 = -i$. Taking the fourth roots on both sides.

$z = (-i)^{1\ 4} = \underline{/(3\pi/2 + 2k\pi)/4}$ by applying De Moivre's theorem

The four roots for $k = 0, 1, 2, 3$ are

$z_1 = \underline{/3\pi/8}$, $z_2 = \underline{/7\pi/8}$, $z_3 = \underline{/11\pi/8}$, $z_4 = \underline{/15\pi/8}$

These roots are shown in the Argand diagram.

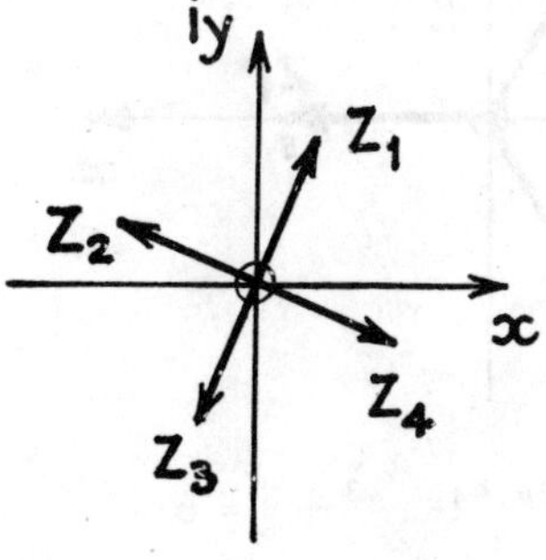

Fig.62 $z_1 = \underline{/3\pi/8}$, $z_2 = \underline{/7\pi/8}$, $z_3 = \underline{/11\pi/8}$, $z_4 = \underline{/15\pi/8}$

(iii) $z^5 = -32$. Taking the fifth roots on both sides.

$z = (-32)^{1/5} = 32^{1/5} \underline{/(\pi + 2k\pi/5}$ by applying De Moivre's theorem

For $k = 0, 1, 2, 3, 4$, the roots are

$z_1 = 2\underline{/\pi/5}$, $z_2 = 2\underline{/3\pi/5}$, $z_3 = 2\underline{/\pi}$, $z_4 = 2\underline{/7\pi/5}$, $z_5 = 2\underline{/9\pi/5}$.

These roots are shown in the Argand diagram.

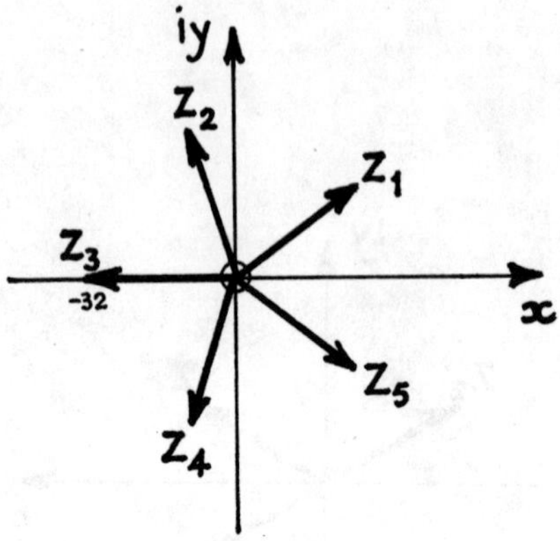

Fig.63 $z^5 + 32 = 0$

(iv) $z^3 = 8$. Taking the cube roots on both sides

$z = 2(1)^{1/3} = 2\underline{/(2k\pi)/3}$ by applying De Moivre's theorem.

The roots are as shown below for $k = 0, 1, 2$

$z_1 = 2\underline{/0^o}$ $z_2 = 2\underline{/2\pi/3}$ $z_3 = 2\underline{/4\pi/3}$.

These roots are shown in the Argand diagram.

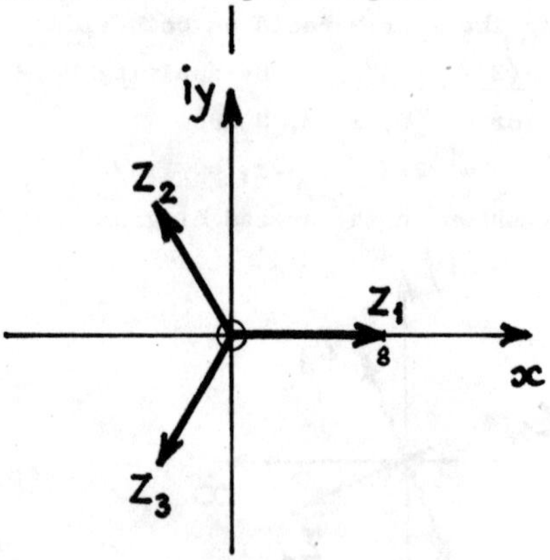

Fig.64 $z^3 - 8 = 0$

4 (i) $(z + i)^6 + (z - i)^6 = 0$

Expanding each binomial, we have

$$z^6 + 6z^5 i + 15z^4 i^2 + 20z^3 i^3 + 15z^2 i^4 + 6z i^5 + i^6$$
$$+ z^6 - 6z^5 i + 15z^4 i^2 - 20z^3 i^3 + 15z^2 i^4 - 6z i^5 + i^6 = 0$$

This is simplified

$$2z^6 + 30z^4 i^2 + 15z^2 i^4 + 2i^6 = 0$$
$$z^6 - 15z^4 + 15z^2 - 1 = 0$$

Let $w = z^2$

$$w^3 - 15w^2 + 15w - 1 = 0 \quad \text{..........................(1)}$$

Solving this cubic equation by trial and error method first,

Let $w = 1$

$$1 - 15 + 15 - 1 = 0$$

therefore $w - 1$ is a factor.

To find the other factors, divide equation (1) by the factor

$$\begin{array}{r|l} & w^2 - 14w + 1 \\ \hline w - 1 & w^3 - 15w^2 + 15w - 1 \\ & w^3 - w^2 \\ \hline & \quad - 14w^2 + 15w - 1 \\ & \quad - 14w^2 + 14w \\ \hline & \qquad\qquad w - 1 \\ & \qquad\qquad w - 1 \\ \hline & \qquad\qquad - \quad - \end{array}$$

The cubic equation is factorised as follows:

$$w^3 - 15w^2 + 15w - 1 = (w - 1)(w^2 - 14w + 1) = 0$$
$$w = z^2 = 1 \quad \text{and} \quad w^2 - 14w + 1 = 0$$

Therefore, $z = \pm 1$, and $w = \dfrac{14 \pm \sqrt{196 - 4}}{2} = 7 \pm 4\sqrt{3}$

$w = 7 + 4\sqrt{3} = z^2$ $\qquad z = \pm (7 + 4\sqrt{3})^{\frac{1}{2}} = \pm 13.93^{\frac{1}{2}}$

$w = 7 - 4\sqrt{3} = z^2$ $\qquad z = \pm (7 - 4\sqrt{3})^{\frac{1}{2}} = \pm 0.267$

$z = \pm 1, \ \pm 13.93^{\frac{1}{2}}, \ \pm 0.267$ $\qquad z = \pm 1, \ \pm 3.732, \ \pm 0.267$

The six roots of the initial equation are

$z_1 = 1$ $\qquad z_4 = -3.732$

$z_2 = -1$ $\qquad z_5 = 0.267$

$z_3 = 3.732$ $\qquad z_6 = -0.267$

which are all real roots.

Alternatively

$$(z + i)^6 + (z - i)^6 = 0$$

$$(z + i)^6 = -(z - i)^6 = (z - i)^6 (-1)$$

Taking the sixth root on each side

$(z + i) = (z - i)(-1)^{1/6}$ where $(-1)^{1/6} = \underline{/(\pi + 2k\pi)}^{1/6}$ where $k = 0,1,2....5$

$$(z + i) = (z - i)\left[\cos\frac{(2k + 1)\pi}{6} + i \sin\frac{(2k + 1)\pi}{6}\right]$$

$$z\left[1 - \cos\frac{(2k + 1)\pi}{6} - i \sin\frac{(2k + 1)\pi}{6}\right] = -i\left[1 + \cos\frac{(2k + 1)\pi}{6} + i \sin\frac{(2k + 1)\pi}{6}\right]$$

$$= \frac{\sin(2k + 1)\pi}{6} - i\left[\cos\frac{(2k + 1)\pi}{6} + 1\right]$$

Employing $\cos 2\theta = 1 - 2\sin^2\theta = 2\cos^2\theta - 1$

$1 - \cos 2\theta = 2\sin^2\theta$ and $\sin 2\theta = 2 \sin\theta \cos\theta$

$$z\left[2 \sin^2\frac{(2k + 1)\pi}{12} - i\ 2 \sin\frac{(2k + 1)\pi}{12} \cos\frac{(2k + 1)\pi}{12}\right]$$

$$= 2 \sin\frac{(2k + 1)\pi}{12} \cos\frac{(2k + 1)\pi}{12} - i\ 2\cos^2\frac{(2k + 1)\pi}{12}$$

$$2 \sin\frac{(2k + 1)\pi}{12}\ Z\left[\sin\frac{(2k + 1)\pi}{12} - i \cos\frac{(2k + 1)\pi}{12}\right]$$

$$= 2 \cos\frac{(2k + 1)\pi}{12}\left[\sin\frac{(2k + 1)\pi}{12} - i \cos\frac{(2k + 1)\pi}{12}\right]$$

therefore $Z = \cos\frac{(2k + 1)\pi}{12} \bigg/ \sin\frac{(2k + 1)\pi}{12}$

or $Z = \cot\frac{(2k + 1)\pi}{12}$ where $k = 0, 1, 2, 3, 4, 5.$

The six roots are

$Z_1 = \cot \pi/_{12} = 3.732$	$Z_4 = \cot 7\pi/_{12} = -0.268$
$Z_2 = \cot \pi/_4 = 1.000$	$Z_5 = \cot 9\pi/_{12} = -1.000$
$Z_3 = \cot 5\pi/_{12} = 0.268$	$Z_6 = \cot 11\pi/_{12} = 3.732$

and these agree with the six roots we found in the first method.

(ii) Show that the roots of the equation $(z + 1)^6 + (z - 1)^6 = 0$ may be expressed in the form

$-i \cot (2r + 1)\pi/12$ where $r = 1, 2, 3, 4, 5, 6$

SOLUTION

Expanding $(z + 1)^6$ and $(z - 1)^6$

$$z^6 + 6z^5 + 15z^4 + 20z^3 + 15z^2 + 6z + 1 + z^6 - 6z^5 + 15z^9 - 6z + 1 = 0$$
$$2z^6 + 30z^4 + 30z^2 + 2 = 0$$
$$z^6 + 15z^4 + 15z^2 + 1 = 0$$

For $r = 1$, one root will be $-i \cot 3\pi/12 = -i$
For $r = 4$, another root will be $-i \cot 9\pi/12 = i$

Let $z^2 = w$

$$w^3 + 15w^2 + 15w + 1 = 0$$

If $w = -1 = z^2$ then $z = \pm\sqrt{-1} = \pm i$

$$w^3 + 15w^2 + 15w + 1 = (w + 1) . (w^2 + bw + c)$$
$$= w^3 + w^2 + bw^2 + bw + cw + c$$
$$= w^3 + w^2 (b + 1) + w (b + c) + c$$

Equating coefficients

$$b + 1 = 15 \quad b = 14 \qquad b + c = 15 \quad c = 1$$

$$w^3 + 15w^2 + 15w + 1 = (w + 1) (w^2 + 14w + 1) = 0$$
$$w^2 + 14w + 1 = 0$$

therefore, $w = \dfrac{-14 \pm \sqrt{14^2 - 4}}{2} = -7 \pm 4\sqrt{3}$

$w = -7 + 4\sqrt{3} = z^2 \quad \therefore \quad z = (-7 + 4\sqrt{3})^{\frac{1}{2}} = 0.268(-1)^{\frac{1}{2}}$

$w = -7 - 4\sqrt{3} = z^2 \quad\quad z = (-7 - 4\sqrt{3})^{\frac{1}{2}} = 3.732(-1)^{\frac{1}{2}}$

$z_1 = i = i \cot \pi/4$ where $\cot \pi/4 = 1$

$z_2 = -i = -i \cot \pi/4$

$z_3 = +i0.268 = i \cot 5\pi/12$

$z_4 = -i0.268 = i \cot 5\pi/12$

$z_5 = +i3.732 = -i \cot 11\pi/12$

$z_6 = -i3.732 = -i \cot 11\pi/12$

$$-i \cot (2r + 1)\pi/12, \quad r = 1, 2, 3, 4, 5, 6$$

Alternatively,

$(z + 1)^6 = - (z - 1)^6$ or $(z + 1) = (z - 1)(-1)^{1/6} = (z - 1)\underline{/(\pi + 2k\pi)}^{1/6}$
where $k = 0, 1, 2, 3, 4, 5$

$$z + 1 = (z - 1) \left(\cos \frac{(2k + 1)\pi}{6} + i \sin \frac{(2k + 1)\pi}{6}\right)$$

$$z\left(1 - \cos \frac{(2k + 1)\pi}{6} - i \sin \frac{(2k + 1)\pi}{6}\right) = -\cos \frac{(2k + 1)\pi}{6} - i \sin\frac{(2k + 1)\pi}{6} - 1$$

$$z\left(2 \sin^2 \frac{(2k+1)\pi}{12} - 2i \sin\frac{(2k+1)\pi}{12} \cos \frac{(2k+1)\pi}{12}\right)$$
$$= -\left(1 + \cos \frac{(2k+1)\pi}{6} + \sin \frac{(2k+1)\pi}{6}\right)$$
$$= -\left(2 \cos \frac{(2k + 1)\pi}{12} + 2i \sin \frac{(2k + 1)\pi}{12} . \cos \frac{(2k + 1)\pi}{12}\right)$$

$$2z \sin \frac{(2k+1)\pi}{12}\left(\sin\frac{(2k+1)\pi}{12} - \cos \frac{(2k+1)\pi}{12}\right) =$$
$$= -2 \cos \frac{(2k+1)\pi}{12}\left(\cos \frac{(2k + 1)\pi}{12} + \sin \frac{(2k+1)\pi}{12}\right)$$

therefore $\quad z = -i \cot \dfrac{(2k + 1)\pi}{12}.$

5. $(1 + z)^n = (1 - z)^n$

Taking the nth root on both sides

$$(1 + z) - (1 - z)\,(1)^{1/n} = (1 - z)\left(\cos 0 + i \sin 0\right)^{1/n}$$

$$(1 + z) = (1 - z)\left(\cos \frac{2k\pi}{n} + i \sin \frac{2k\pi}{n}\right)$$

$$= \cos \frac{2k\pi}{n} + i \sin \frac{2k\pi}{n} - z \cos \frac{2k\pi}{n} - zi \sin \frac{2k\pi}{n}$$

$$z + z \cos \frac{2k\pi}{n} + zi \sin \frac{2k\pi}{n} = -1 + \cos \frac{2k\pi}{n} + i \sin \frac{2k\pi}{n}$$

$$\therefore \quad z\left(1 + \cos \frac{2k\pi}{n} + i \sin \frac{2k\pi}{n}\right) = -1 + \cos \frac{2k\pi}{n} + i \sin \frac{2k\pi}{n}$$

but $\cos 2\theta - 2 \cos^2\theta - 1 = 1 - 2 \sin^2\theta$ and $\sin 2\theta = 2 \sin \theta \cos \theta$

$$z\left(2 \cos \frac{2k\pi}{2n} + 2 \sin \frac{2k\pi}{2n} \cos \frac{2k\pi}{2n}\right) = -2 \sin \frac{2k\pi}{2n} + 2 \sin \frac{2k\pi}{2n} \cos \frac{2k\pi}{2n}$$

$$z2 \cos \frac{k\pi}{n}\left(\cos \frac{k\pi}{n} + i \sin \frac{k\pi}{n}\right) = -2 \sin \frac{k\pi}{n}\left(\sin \frac{k\pi}{n} - i \cos \frac{k\pi}{n}\right)$$

$$z = -\tan \frac{k\pi}{n} \, \frac{\sin \frac{k\pi}{n} - i \cos \frac{k\pi}{n}}{\cos \frac{k\pi}{n} + \sin \frac{k\pi}{n}}$$

Multiplying both sides by i

$$iz = -\tan \frac{k\pi}{n} \left(\frac{i \sin \frac{k\pi}{n} + \cos \frac{k\pi}{n}}{\cos \frac{k}{n} + i \sin \frac{k}{n}}\right)$$

Multiplying both sides by $-i$

$$- i z = + i \tan \frac{k\pi}{n}$$

$$z = i \tan \frac{k\pi}{n}$$

$k = 0, 1, 2, \ldots\ldots (n - 1).$

6 (i) $c(8, 9)$ the coordinates of the centre and $r = 7$ the radius.

$$(x - 8)^2 + (y - 9)^2 = 7^2$$
$$(x - 8) + i\,(y - 9) = 7$$
$$x + iy - 8 - 9i = 7$$

The complex number equation is therefore

$$|\, z - 8 - 9i \,| < 7$$

(ii) $c(a, b)$ the coordinates of the centre and $r = c$

$$(x - a)^2 + (y - b)^2 = c^2$$
$$|x - a + i\,(y - b)| = c$$
$$|x + iy - a - ib| = c$$

The complex number equation is therefore

$$|z - a - ib| = c$$

(iii) $c\,(-1, 0)$ $r = 1$ the coordinates of the centre and radius respectively.

$$(x + 1)^2 + 1^2 = 1^2$$
$$x + 1 + iy = 1$$

The complex number equation is therefore

$$\boxed{|\, z + 1| > 1}$$

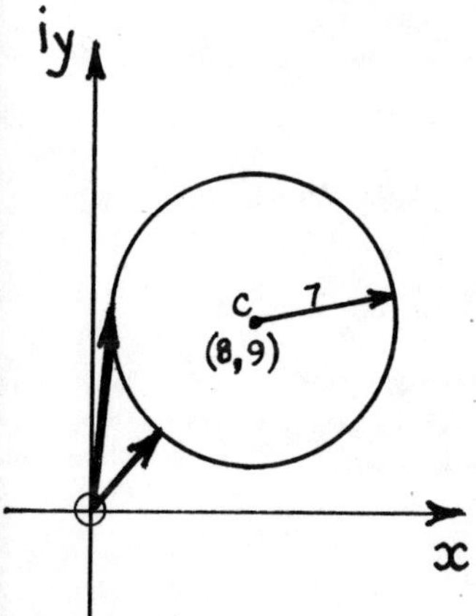

Fig.65 $|z - 8 - 9i| \leq 7$

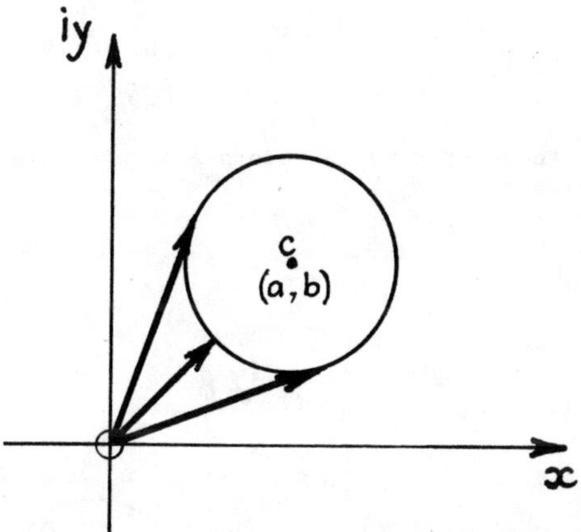

Fig 66 $|z - a - ib| = c$

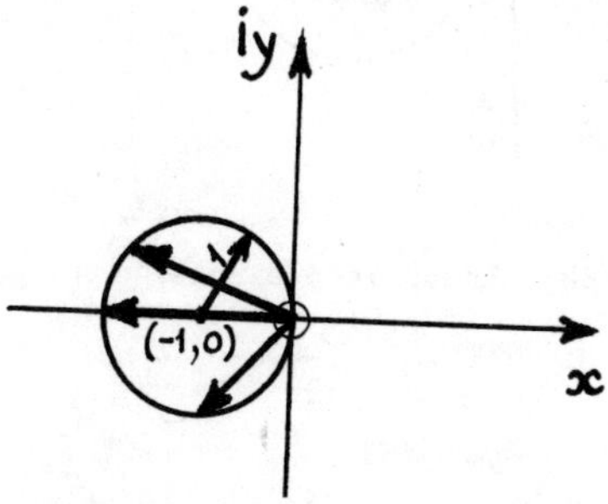

Fig.67 $|z + 1| > 1$

7. It is required to find the locus of z, where
$$\left|\frac{z-1}{z+2}\right| = \frac{1}{2}$$

Let $z = x + iy$

$$\left|\frac{z-1}{z+1}\right| = \left|\frac{x+iy-1}{x+iy+1}\right| = \left|\frac{(x-1)+iy}{(x+1)+iy}\right| = \left|\frac{(x-1)+iy}{(x+1)+iy}\right|$$

therefore $\dfrac{\sqrt{(x-1)^2+y^2}}{\sqrt{(x+1)^2+y^2}} = \dfrac{1}{2}$ ....(1)

Squaring up both sides of equation (1)

$$\frac{(x-1)^2+y^2}{(x+1)^2+y^2} = \frac{1}{4}$$

$$4(x^2 - x + 1 + y^2) = x^2 + 2x + 1 + y^2$$

$$3x^2 - 10x + 3 + 3y^2 = 0$$

$$3x^2 + 3y^2 - 10x + 3 = 0$$

Dividing each term by 3

$$x^2 + y^2 - \frac{10x}{3} + 1 = 0$$

Completing the squares

$$\left(x - \frac{5}{3}\right)^2 - \left(\frac{5}{3}\right)^2 + y^2 + 1 = 0$$

$$\left(x - \frac{5}{3}\right)^2 + y^2 = \frac{25}{9} - \frac{9}{9}$$

$$\left(x - \frac{5}{3}\right)^2 + y^2 = \left(\frac{4}{3}\right)^2$$

The locus of z is therefore a circle with centre $c(^5/_3, 0)$ and radius $r = {}^4/_3$.

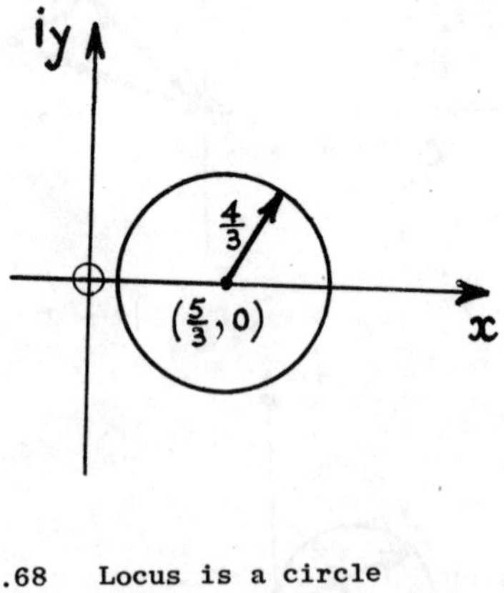

Fig.68   Locus is a circle
$(5/_3, 0)$, $r = 4/_3$

8 SOLUTIONS

(i) It is required to find the cartesian equation of the locus of P when

$$\left|\frac{z + 1}{z + 2i}\right| = 5$$

Let $z = x + iy$

$$\frac{|x + iy + 1|}{|x + iy + 2i|} = \left|\frac{(x + 1) + iy}{x + i(y + 2)}\right| = \frac{\sqrt{(x + 1)^2 + y^2}}{\sqrt{x^2 + (y + 2)^2}} = 5$$

therefore $\dfrac{\sqrt{(x + 1)^2 + y^2}}{\sqrt{x^2 + (y + 2)^2}} = 5$

Squaring up both sides

$$\frac{(x + 1)^2 + y^2}{x^2 + (y + 2)^2} = 5^2$$

$$x^2 + 2x + 1 + y^2 = 25x^2 + 25y^2 + 100y + 100$$

$$24x^2 + 24y^2 - 2x - 1 + 100y + 100 = 0$$

the cartesian equation which is a circle.

Dividing each term by 24

$$x^2 - \frac{1}{12}x + y^2 + \frac{100y}{24} + \frac{100}{24} - \frac{1}{24} = 0$$

Completing the squares

$$\left(x - \frac{1}{24}\right)^2 - \frac{1}{24^2} + \left(y + \frac{50}{24}\right)^2 - \left(\frac{50}{24}\right)^2 + \frac{100}{24} - \frac{1}{24} = 0$$

$$\left(x - \frac{1}{24}\right)^2 + \left(y + \frac{25}{12}\right)^2 = \frac{1}{576} + \frac{2500}{576} - \frac{99}{24}$$

$$= \frac{2501 - 99 \times 24}{576}$$

$$= \frac{125}{576} = \left(\frac{5\sqrt{5}}{24}\right)^2$$

This is a circle with the centre having coordinates $C\left(\frac{1}{24}, -\frac{25}{12}\right)$ and radius $= \dfrac{5\sqrt{5}}{24}$

9. THE FIFTH ROOTS OF -1

To solve $z^5 + 1 = 0$
$z^5 = -1$

$(\cos\theta + i\sin\theta)^5 = -1 = \cos\pi + i\sin\pi$

Applying De Moivre's theorem

$\cos 5\theta + i\sin 5\theta = (\cos\pi + i\sin\pi) = \cos(\pi + 2k\pi) + i\sin(\pi + 2k\pi)$

Equating real and imaginary terms

$5\theta = \pi + 2\pi$ and $\theta = \dfrac{(2k+1)\pi}{5}$ where $k = 0, 1, 2, 3, 4$

then $\theta = \pi/5,\quad 3\pi/5,\quad \pi,\quad 7\pi/5,\quad 9\pi/5.$

The complex fifth roots of -1 occur in conjugate pairs.

$z_1 = \cos\pi/5 + i\sin\pi/5$

$z_2 = \cos 3\pi/5 + i\sin 3\pi/5$

$z_3 = \cos\pi + i\sin\pi$

$z_4 = \cos\dfrac{7\pi}{5} + i\sin\dfrac{7\pi}{5} = \cos(-3\pi/5) + i\sin(-3\pi/5)$

$z_5 = \cos\dfrac{9\pi}{5} + i\sin\dfrac{9\pi}{5} = \cos(-\pi/5) + i\sin(-\pi/5)$

The sum of the complex nth roots of any number is zero.

$z_1 + z_2 + z_3 + z_4 + z_5 = 0$

$Re\,(z_1 + z_2 + z_3 + z_4 + z_5) = 0$
$Im\,(z_1 + z_2 + z_3 + z_4 + z_5) = 0$

$\cos\pi/5 + \cos 3\pi/5 + (-1) + \cos(-\pi/5) + \cos(-3\pi/5) = 0$
$2\cos\pi/5 + 2\cos 3\pi/5 = 1$

$$\cos\pi/5 + \cos 3\pi/5 = \frac{1}{2}$$

10 If $z^9 - 1 = 0$, show $\cos 2\pi/9 + \cos 4\pi/9 + \cos 6\pi/9 + \cos 8\pi/9 = -\dfrac{1}{2}$

SOLUTION

$z = 1^{1/9} = \underline{/2k\pi/9}$, $k = 0, \pm1, \pm2, \pm3, \pm4$

$z_1 = 1$ $z_2 = \underline{/2\pi/9}$ $z_3 = \underline{/-2\pi/9}$

$z_4 = \underline{/4\pi/9}$ $z_5 = \underline{/-4\pi/9}$ $z_6 = \underline{/6\pi/9}$

$z_7 = \underline{/-6\pi/9}$ $z_8 = \underline{/8\pi/9}$ $z_9 = \underline{/-8\pi/9}$

$Re\,(z_1 + z_2 + \ldots\ldots\ldots + z_9) = 0$

$1 + \cos 2\pi/9 + \cos(-2\pi/9) + \cos 4\pi/9 + \cos(-4\pi/9)$
$+ \ldots\ldots\ldots + \cos 8\pi/9 + \cos(-8\pi/9) = 0$

$Im(z_1 + z_2 + \ldots\ldots\ldots + z_9) = 0$

$1 + 2\cos 2\pi/9 + 2\cos 4\pi/9 + 2\cos 6\pi/9 + 2\cos 8\pi/9 = 0$

therefore $\cos 2\pi/9 + \cos 4\pi/9 + \cos 6\pi/9 + \cos 8\pi/9 = -\dfrac{1}{2}$

11.. If $z^7 + 1 = 0$, show $\cos \pi/7 + \cos 3\pi/7 + \cos 5\pi/7 = \frac{1}{2}$

SOLUTION

$z = (-1)^{1/7} = \underline{/(\pi + 2k\pi)}^{1/7} = \underline{/(2k + 1)}\pi/7$

$z_1 = \cos \pi/7 + i \sin \pi/7$ $\qquad k = 0, \pm 1, \pm 2, \pm 3$

$z_2 = \cos 3\pi/7 + i \sin 3\pi/7$

$z_3 = \cos 5\pi/7 + i \sin 3\pi/7$

$z_4 = \cos 7\pi/7 + i \sin 7\pi/7$

$z_5 = \cos (-\pi/7) + i \sin (-\pi/7)$

$z_6 = \cos(- 3\pi/7) + i \sin (-3\pi/7)$

$z_7 = \cos (-5\pi/7) + i \sin (-5\pi/7)$

$Re(z_1 + z_2 + \ldots\ldots\ldots + z_7) = 2 \cos \pi/7 + 2 \cos 3\pi/7 + 2 \cos 5\pi/7 + \cos 7\pi/7 = 0$

Therefore, $\cos \pi/7 + \cos 3\pi/7 + \cos 5\pi/7 = \frac{1}{2}$

12. Find the six roots of the equation with real coefficients

$$z^6 - 2z^3 + 4 = 0$$

SOLUTIONS

Let $z^3 = w$

$w^2 - 2w + 4 = 0 \qquad w = \frac{2 \pm \sqrt{4 - 16}}{2} = 1 \pm i\sqrt{3}$

$w = z^3 = 1 + i\sqrt{3}$

OR $z = (1 + i\sqrt{3})^{1/3} = (2\underline{/\pi/3})^{1/3} = 2^{1/3} \underline{/(\pi/3 + 2k\pi)/3}$

$z_1 = 2^{1/3} \underline{/\pi/9} = 2^{1/3} (\cos \pi/9 + i \sin \pi/9)$ when $k = 0$

$z_2 = 2^{1/3} \underline{/7\pi/9} = 2^{1/3} (\cos 7\pi/9 + i \sin 7\pi/9)$ when $k = 1$

$z_3 = 2^{1/3} \underline{/13\pi/9} = 2^{1/3} (\cos 13\pi/9 + i \sin 13\pi/9)$ when $k = 2$

and $w = z^3 = 1 - i\sqrt{3}$

Since the roots occur in conjugate pairs, then

$z_4 = \bar{z}_1 = 2^{1/3} \underline{/-\pi/9} = 2^{1/3} (\cos \pi/9 - i \sin \pi/9)$

$z_5 = \bar{z}_2 = 2^{1/3} \underline{/-7\pi/9} = 2^{1/3} (\cos 7\pi/9 - i \sin 7\pi/9)$

$z_6 = \bar{z}_3 = 2^{1/3} \underline{/-13\pi/9} = 2^{1/3} (\cos 13\pi/9 - i \sin 13\pi/9)$

Find the product of three quadratic factors with real coefficients.

$(z - z_1)(z - \bar{z}_1) = z^2 - z(z_1 + \bar{z}_1) + z_1\bar{z}_1 = z^2 - z\, 2^{4/3} \cos \pi/9 + 2^{2/3}$

$(z - z_2)(z - \bar{z}_2) = z^2 - z(z_2 + \bar{z}_2) + z_2\bar{z}_2 = z^2 - z\, 2^{4/3} \cos 7\pi/9 + 2^{2/3}$

$(z - z_3)(z - \bar{z}_3) = z^2 - z(z_3 + \bar{z}_3) + z_3\bar{z}_3 = z^2 - z\, 2^{4/3} \cos 13\pi/9 + 2^{2/3}$

13. <u>**SOLUTION**</u>

$z^6 - 1 = 0 \quad \ldots\ldots\ldots\ldots\ldots\ldots\ldots\ldots\ldots\ldots (1)$

$(z - 1)(z^5 + z^4 + z^3 + z^2 + z + 1) = 0$

The six roots of (1)

$z = 1^{1/6} = (\cos 0 + i \sin 0)^{1/6} = \underline{/2k\pi/6}$ where $k = 0, 1, 2, 3, 4, 5$

$z_1 = \underline{/0^{\circ}} = \cos 0 + i \sin 0 = 1$

$z_2 = \underline{/\pi/3} = \cos \pi/3 + i \sin \pi/3 = \frac{1}{2} + i\frac{\sqrt{3}}{2} = \bar{z}_6 = e^{i\pi/3}$

$z_3 = \underline{/2\pi/3} = \cos 2\pi/3 + i \sin 2\pi/3 = -\frac{1}{2} + i\frac{\sqrt{3}}{2} = \bar{z}_5 = -e^{-i2\pi/3}$

$z_4 = \underline{/\pi} = \cos \pi + i \sin \pi = -1$

$z_5 = \underline{/4\pi/3} = \cos 4\pi/3 + i \sin 4\pi/3 = -\frac{1}{2} - i\frac{\sqrt{3}}{2} = \bar{z}_3 = -e^{i4\pi/3} = -e^{-i2\pi}\ {}_3$

$z_6 = \underline{/5\pi/3} = \cos 5\pi/3 + i \sin 5\pi/3 = \frac{1}{2} - i\frac{\sqrt{3}}{2} = \bar{z}_2 = e^{-i5\pi/3} = e^{-i\pi/3}$

$z = \pm 1, \quad \pm e^{\pm i\pi/3}$

The five roots of $z^5 + z^4 + z^3 + z^2 + z + 1 = 0$
are $-1, \quad e^{\pm i\pi/3}, \quad e^{\pm i2\pi/3}$.

14. $z^4 + 1 = (z^2 - i)(z^2 + i) = 0$

$z^4 = -1, \quad z = (-1)^{1/4} = (\cos \pi + i \sin \pi)^{1/4}$

$z = (-1)^{1/4} = \underline{/(\pi + 2k\pi)/4} = \cos \pi \dfrac{(1 + 2k)}{4} + i \sin \pi \dfrac{(1 + 2k)}{4}$

where $k = 0, 1, 2, 3$

$z_1 = \underline{/\pi/_4} = \cos \pi/_4 + i \sin \pi/_4 = \dfrac{1}{\sqrt{2}} + i\dfrac{1}{\sqrt{2}}$ when $k = 0$

$z_2 = \underline{/3\pi/_4} = \cos 3\pi/_4 + i \sin 3\pi/_4 = -\dfrac{1}{\sqrt{2}} + i\dfrac{1}{\sqrt{2}}$ when $k = 1$

$z_3 = \underline{/5\pi/_4} = \cos 5\pi/_4 + i \sin 5\pi/_4 = -\dfrac{1}{\sqrt{2}} - i\dfrac{1}{\sqrt{2}}$ when $k = 2$

$z_4 = \underline{/7\pi/_4} = \cos 7\pi/_4 + i \sin 7\pi/_4 = \dfrac{1}{\sqrt{2}} - i\dfrac{1}{\sqrt{2}}$ when $k = 3$

$z_2 = \overline{z}_3$ or $z_3 = \overline{z}_2$ and $z_1 = \overline{z}_4$ or $z_4 = \overline{z}_1$

$$\left[z - \left(\frac{1}{\sqrt{2}} + i\frac{1}{\sqrt{2}}\right)\right]\left[z - \left(\frac{1}{\sqrt{2}} - i\frac{1}{\sqrt{2}}\right)\right]\left[z - \left(-\frac{1}{\sqrt{2}} + i\frac{1}{\sqrt{2}}\right)\right]\left[z - \left(-\frac{1}{\sqrt{2}} - i\frac{1}{\sqrt{2}}\right)\right]$$

$$= \left[z^2 - \left(\frac{1}{\sqrt{2}} + i\frac{1}{\sqrt{2}}\right)^2\right]\left[z^2 - \left(\frac{1}{\sqrt{2}} - i\frac{1}{\sqrt{2}}\right)^2\right]$$

$$= \left[z^2 - \frac{1}{2} - i^2\frac{1}{2} - \frac{2i}{\sqrt{2}\sqrt{2}}\right]\left[z^2 - \frac{1}{2} - i^2\frac{1}{2} + \frac{2i}{\sqrt{2}\sqrt{2}}\right]$$

$= (z^2 - i)(z^2 + i)$

Note that the $Re(z_1 + z_2 + z_3 + z_4) = 0$

and $Im(z_1 + z_1 + z_2 + z_3) = 0$

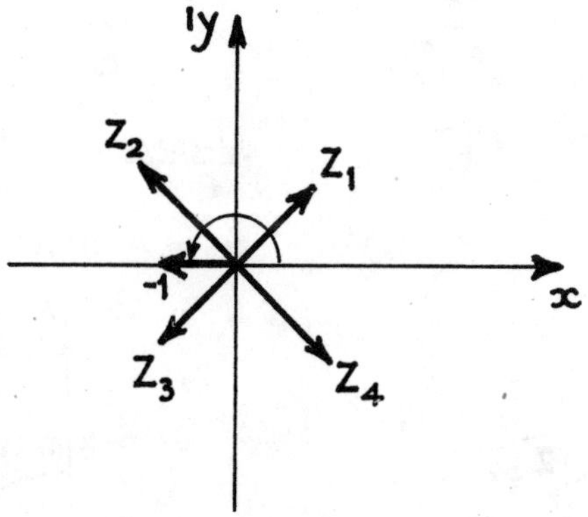

Fig 69 The roots of $z^4 + 1 = 0$.

15. $(z^n - e^{i\theta})(z^n - e^{-i\theta}) = z^{2n} - z^n e^{i\theta} - z^n e^{-i\theta} + e^{i\theta} e^{-i\theta}$

$= z^{2n} - z^n (e^{i\theta} + e^{-i\theta}) + 1$

$= z^{2n} - 2z^n \cos\theta + 1$

since $e^{i\theta} = 1 + \frac{i\theta}{1!} + \frac{i^2\theta^2}{2!} + \frac{i^3\theta^3}{3!} + \frac{i^4\theta^4}{4!} + \ldots..$

$e^{-i\theta} = 1 - \frac{i\theta}{1!} + \frac{i^2\theta^2}{2!} - \frac{i^3\theta^3}{3!} + \frac{i^4\theta^4}{4!}$

$e^{i\theta} + e^{-i\theta} = 2 - 2\frac{\theta^2}{2!} + 2\frac{\theta^4}{4!} - ,\ldots\ldots.$

$= 2\left(1 - \frac{\theta^2}{2!} + \frac{\theta^4}{4!} \ldots\ldots\right) = 2\cos\theta$

If $n = 4$

$(z^4 - e^{i\theta})(z^4 - e^{-i\theta}) = z^8 - 2z^4 \cos\theta + 1$

comparing with $z^8 + z^4\sqrt{3} + 1 = 0$

then $\sqrt{3} = -2\cos\theta$ $\therefore$ $\cos\theta = -\frac{\sqrt{3}}{2}$ $\theta = 5\pi/6$

$(z^4 - e^{i5\pi/6})(z^4 - e^{-i5\pi/6}) = 0$

$z^4 - e^{i5\pi/6} = 0$ $z = (e^{i5\pi/6})^{1/4} = (\cos 5\pi/6 + i\sin 5\pi/6)^{1/4}$

$z = \cos(5\pi/6 + 2k\pi)1/4 + i\sin(5\pi/6 + 2k\pi)1/4$ where $k = 0,1,2,3$

For $k = 0$

$z_1 = \cos 5\pi/24 + i\sin 5\pi/24 = \underline{/5\pi/24} = \underline{/37}.5^o$

For $k = 1$

$z_2 = \underline{/5\pi/24 + \pi/2} = \underline{/127.5}^o$

For $k = 2$

$z_3 = \underline{/5\pi/24 + \pi} = \underline{/217}.5^o$

For $k = 3$

$z_4 = \underline{/5\pi/24 + 3\pi/2} = \underline{/307.5}^o$ or $z_4 - e^{-i5\pi/6} = 0$

$z_4 = e^{-i5\pi/6}$ $z_4 = e^{-i5\pi/6}\ ^{1/4} = \underline{/-(5\pi/6} + 2k\pi)^{1/4}$

For $k = 0, 1, 2, 3$

$z_5 = \underline{/-37}.5,$ $z_6 = \underline{/-127}.5^o,$ $z_7 = \underline{/-217}.5^o$ $z_8 = \underline{/-307}.5$

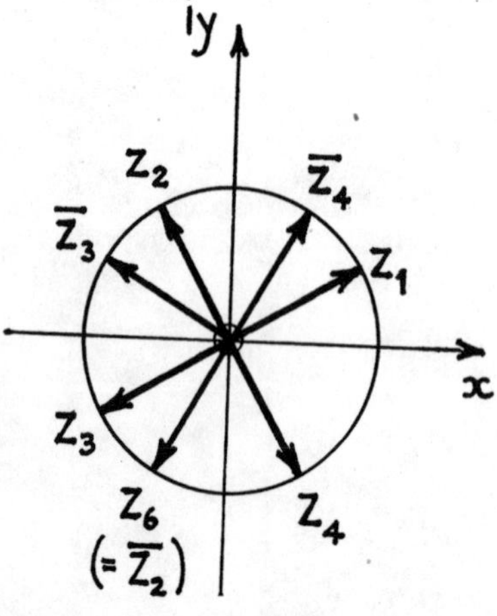

Fig.70

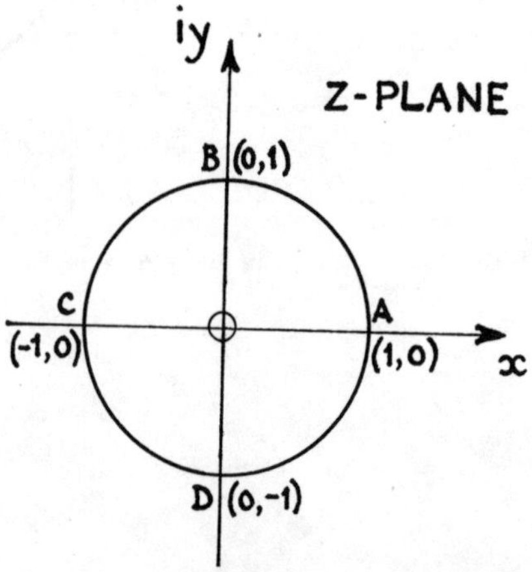

Fig.71

SOLUTIONS 20

1. $w = 2z + 3 = 2(x + iy) + 3 = 2x + 3 + i2y = u + iv$

$u = 2x + 3 \quad \therefore \quad x = \dfrac{u - 3}{2}$

$v = 2y \quad \therefore \quad y = v/2$

$x^2 + y^2 = \left(\dfrac{u - 3}{2}\right)^2 + \left(\dfrac{v}{2}\right)^2 = 1$ since $|z| = 1$, $x^2 + y^2 = 1$

The corresponding motion of q is $(u - 3)^2 + v^2 = 2^2$ which is a circle with centre $C(3, 0)$ and radius $r = 2$.

2. $w = 2 + iz = 2 + i(x + iy) = 2 + ix + i^2y = (2 - y) + ix = u + iv$

Equating real and imaginary terms $u = 2 - y \qquad y = 2 - u$

$v = x \qquad x = v$

Squaring up both sides of these two equations, and adding
$x^2 + y^2 = v^2 + (2 - u)^2 = 1 \quad = \quad (u - 2)^2 + v^2 = 1$

This is also a circle with centre C (2,0) and $r = 1$.

3. $w = 3z^2 = 3(x + iy)^2 = 3(x^2 + 2xyi + i^2y^2)$
$= (3x^2 - 3y^2) + i6xy = u + iv$

Equating real and imaginary terms
$u = 3x^2 - 3y^2$ but $x^2 + y^2 = 1$, $y^2 = 1 - x^2$
$v = 6xy \qquad v = 6x\sqrt{1 - x^2} \quad \therefore \quad v^2 = 36x^2(1 - x^2) = 36x^2 - 36x^4$
$u = 3x^2 - 3(1 - x^2) = 3x^2 - 3 + 3x^2 = 6x^2 - 3$

$\dfrac{(u + 3)}{6} = x^2$ and $v^2 = 36\dfrac{(u + 3)}{6} - 36\dfrac{(u + 3)^2}{36}$

$v^2 = 6u + 18 - u^2 - 6u - 9, \quad v^2 + u^2 = 9$

$u^2 + v^2 = 3^2$

This is a circle C (0,0) $r = 3$

Alternatively if $z = \cos\theta + i\sin\theta$, then $|z| = 1$
$w = 3z^2 = 3(\cos\theta + i\sin\theta)^2 = 3\cos 2\theta + i3\sin 2\theta$
$w = u + iv = 3\cos 2\theta + i3\sin 2\theta$

Equating real and imaginary terms
$u = 3\cos 2\theta$ and $v = 3\sin 2\theta$
and squaring both sides of these equations $\left(\dfrac{u}{3}\right)^2 = \cos^2 2\theta$ and $\left(\dfrac{v}{3}\right)^2 = \sin^2 2\theta$

therefore, $\cos^2 2\theta + \sin^2 2\theta = 1 = \dfrac{u^2}{3^2} + \dfrac{v^2}{3^2} \qquad u^2 + v^2 = 3^2$

As θ varies from 0 to 2π, then 2θ will vary from 0 to 4π, that is, while in the z-plane a circle is described, and in the w-plane will be described twice.

Alternatively $\dfrac{w}{3} = |z|^2 \qquad |w| = 3|z|^2 = 3$
twice a circle, since we are squaring up $|z| = 1$

4. $w = z^3 = (\cos\theta + i\sin\theta)^3 = \cos 3\theta + i\sin 3\theta = u + iv$
where $u = \cos 3\theta$ and $v = \sin 3\theta$

$|w| = |z|^3 = 1 \qquad |w| = 1 \qquad$ a circle C (0,0) $r = 1$
three times as $|z| = 1$ is cubed.

5. $w = z^{-\frac{1}{2}}$ $\quad w^{-2} = (z^{-\frac{1}{2}})^{-2} = z$ $\quad |z| = \frac{1}{|w|^2} = 1$ $\quad |w|^2 = 1$

Two semi-circles of $|w| = 1$

6. $w = z^2 + 2z$ where $z = x + iy$

$w = (x + iy)^2 + 2(x + iy)$

$w = x^2 + i^2y^2 + 2ixy + 2x + 2iy$

$w = x^2 - y^2 + 2x + i\,(2xy + 2y)$

and since $w = u + iv$, equating real and imaginary terms

$u = x^2 - y^2 + 2x = x^2 - (1 - x^2) + 2x$ since $y^2 = 1 - x$

$v = 2xy + 2y = (2x + 2)\sqrt{1 - x^2}$

Squaring up this expression, $v^2 = 4(x + 1)^2\,(1 - x^2)$

$u = x^2 - 1 + x^2 + 2x = 2x^2 + 2x - 1$

$v^2 = 4(x + 1)^2\,(1 - x^2)$

$v^2 = 4(x^2 + 2x + 1)\,(1 - x^2)$

$v^2 = (4x^2 + 8x + 4)\,(1 - x^2) = 4x^2 + 8x + 4 - 4x^4 - 8x^3 - 4x^2$

$v^2 = -4x^4 - 8x^3 + 8x + 4$

$u^2 = (2x^2 + 2x - 1)^2$

$u^2 = 4x^4 + 4x^2 + 1 + 8x^3 - 4x^2 - 4x$ ....(1)

$v^2 = 4x^2 + 8x + 4 - 4x^4 - 8x^3 - 4x^2$ ....(2)

Adding (1) and (2)

$u^2 + v^2 = 4x^4 + 4x^2 + 1 + 8x^3 - 4x^2 - 4x + 4x^2 + 8x + 4 - 4x^4 - 8x^3 - 4x^2$

$= 5 + 4x$

$w = z^2 + 2z$ where $z = \cos\theta + i\sin\theta$ and $|z| = 1$

$= (\cos\theta + i\sin\theta)^2 + 2(\cos\theta + i\sin\theta)$

$= \cos 2\theta + i\sin 2\theta + 2\cos\theta + 2i\sin\theta)$

$= (\cos 2\theta + 2\cos\theta) + i(\sin 2\theta + 2\sin\theta)$

$= u + iv$

Equating real and imaginary terms

$u = \cos 2\theta + 2\cos\theta$

$v = \sin 2\theta + 2\sin\theta$

Squaring up both sides and adding

$u^2 = \cos^2 2\theta + 4\cos^2\theta + 4\cos\theta\cos 2\theta$

$v^2 = \sin^2 2\theta + 4\sin^2\theta + 4\sin\theta\sin 2\theta$

$u^2 + v^2 = 1 + 4 + 4\,(\sin\theta\sin 2\theta + \cos\theta\cos 2\theta)$

$= 5 + 4\,(\cos 2\theta\cos\theta + \sin 2\theta\sin\theta)$

$u^2 + v^2 = 5 + 4\cos(2\theta - \theta)$

$u^2 + v^2 = 5 + 4\cos\theta$

$r^2 = 1 + 4 + 4\cos\theta$

$r^2 = 1 + 4\,(1 + \cos\theta)$

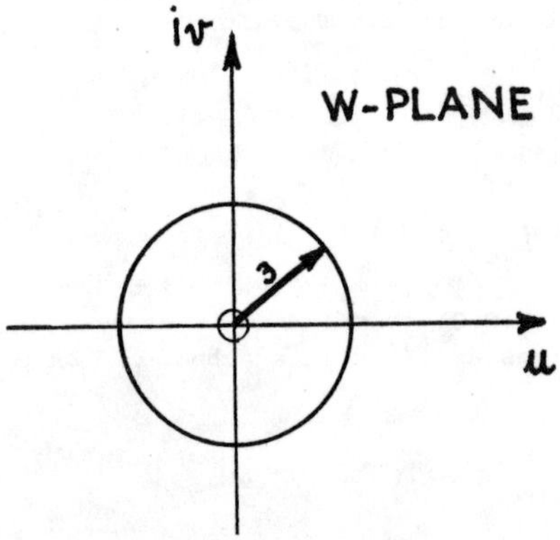

Fig.72 Transformation.

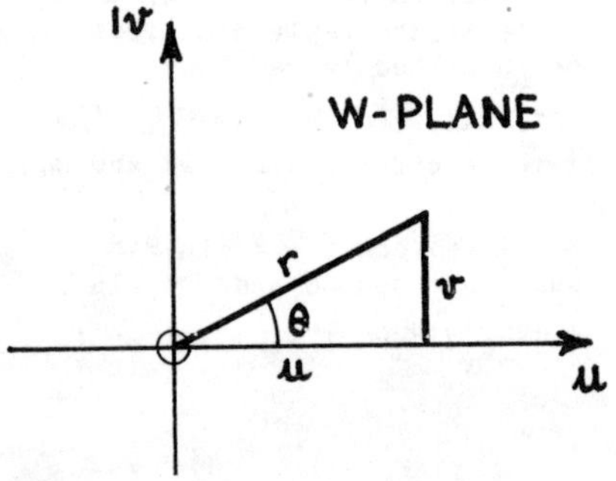

Fig.73 $\cos\theta = u/r$

$\sin\theta = v/r$

7. $w = (z + 1)^{\frac{1}{2}}$, squaring up both sides and substituting $z = x + iy$

$w^2 = (x + iy + 1)$ but $w = u + iv$

$(u + iv)^2 = x + iy + 1$

$(u + iv)^2 = (x + 1) + iy$, expanding the left hand side

$$(u^2 - v^2) + i2uv = (x + 1) + iy$$

and equating real and imaginary parts $x + 1 = u^2 - v^2$,

hence $x = u^2 - v^2 - 1$

$2uv = y$ *OR* $y = 2uv$ *but* $x^2 + y^2 = 1$

then $(u^2 - v^2 - 1)^2 + (2\,uv)^2 = 1$

$u^4 + v^4 + 1 - 2u^2v^2 - 2u^2 + 2v^2 + 4u^2v^2 = 1$

$u^4 + v^4 + 1 + 2u^2v^2 - 2u^2 + 2v^2 = 1$

$(u^2 + v^2)^2 + 1 - 2u^2 + 2v^2 = 1$

$(u^2 + v^2)^2 = 2u^2 - 2v^2$

$(r^2)^2 = 2(r^2 \cos^2\theta - r^2 \sin^2\theta)$

$(r^2)^2 = 2r^2 (\cos^2\theta - \sin^2\theta) = 2r^2 \cos 2\theta$

$$r^2 = 2 \cos 2\theta$$

This polar equation represents a lemniscate i.e. the path of a point $P\ (r, \theta)$, that moves in such a way that the product of its distances from the two points $P_1\ (a, \pi)$ and $P_2\ (a, 0)$ is a constant a^2. In this case $a = 1$.

LEMNISCATE

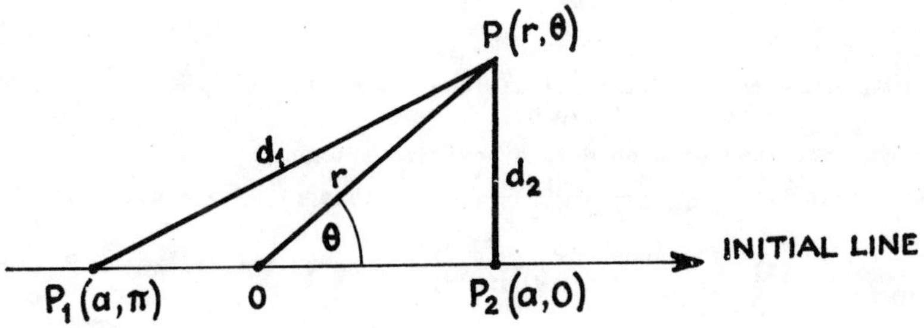

Fig 74 Lemniscale $r^2 = 2a^2\cos 2\theta$.

$PP_1 . PP_2 = k^2$ (a positive constant)

$\Delta\ OPP_2$

$d_2{}^2 = r^2 + a^2 - 2a\,r \cos\theta$

$\Delta\ OPP_1$

$d_1{}^2 = r^2 + a^2 - 2a\,r \cos(\pi - \theta)$

$= r^2 + a^2 + 2a\,r \cos\theta$

$d_1{}^2\, d_2{}^2 = \left[(r^2 + a^2) + 2a\,r \cos\theta\right]\left[(r^2 + a^2) - 2a\,r \cos\theta\right]$

$= (r^2 + a^2)^2 - 4a^2 r^2 \cos^2\theta$

$k^4 = a^4 + r^4 + 2a^2 r^2 - 4a^2 r^2 \cos\theta = a^4 + r^4 + 2a^2 r^2 (1 - 2\cos^2\theta)$

$= a^4 + r^4 - 2a^2 r^2 \cos 2\theta$

If $k^4 = a^4$

$$r^2 = 2a^2 \cos 2\theta$$

EXERCISES

1 (a) By first eliminating z_2, find the complex numbers z_1, z_2 that satisfy the simultaneous equations

$$(1 + i)z_1 - iz_2 = 3 - 2i$$
$$(2 - i)z_1 + (1 - i)z_2 = 4 - 4i.$$

(b) Show that the locus of z defined by the equation

$$z\bar{z} + (2 - i)z + (2 + i)\bar{z} + 1 = 0$$

is a circle.

Find the complex number corresponding to its centre, and find its radius.

(c) Show, in the Argand diagram, the lines defined by the following equations:

$$\arg(z + i) = \tfrac{1}{4}\pi$$
$$Im(\bar{z}) = -1.$$

Find the complex number corresponding to their point of intersection, expressing it both in Cartesian and polar form. W.J.E.C.

Ans. 1(a) $z_1 = 1 - i$, $z_2 = 2 + i$. (c) $z = 2 + i$, $z = \sqrt{5}\,/\tan^{-1}\tfrac{1}{2}$

(b) $|z + 2 + i| = 2$

2 Show that the roots of the equation $5z^4 - 10z^3 + 10z^2 - 5z + 1 = 0$ are $\tfrac{1}{2}\{1 \pm i\cot(\tfrac{1}{5}r\pi)\}$ for $r = 1, 2$. C

3 De Moivre's theorem states that $(\cos\theta + i\sin\theta)^n = \cos n\theta + i\sin n\theta$.

Prove this theorem when n is a positive integer.

Show that (i) $\dfrac{(\cos\theta + i\sin\theta)^4}{(\sin\phi + i\cos\phi)^3} = i\cos(4\theta + 3\phi) - \sin(4\theta + 3\phi)$,

(ii) $\dfrac{(1 + \cos\theta + i\sin\theta)^4}{(1 + \cos\theta - i\sin\theta)^4} = \cos 4\theta + i\sin 4\theta$ O

4 (a) Show that any complex number $z = x + iy$ can be expressed in polar form $z = r(\cos\theta + i\sin\theta)$. Hence prove that, for any two complex numbers z_1, z_2

$$|z_1 z_2| = |z_1| \,.\, |z_2|.$$

Verify that $\arg\left(\dfrac{z_1}{z_2}\right) = \arg z_1 - \arg z_2$ when $z_1 = -\sqrt{3} + i$ and $z_2 = 1 + i\sqrt{3}$

(b) Find the cartesian equation for the locus of points satisfying $Im(z^2) = -2$.

(c) Sketch the region in the Argand plane enclosed by parts of the following four loci:

$$|z| = 4, \quad |z + 1| = 2, \quad \arg z = \pi, \quad \arg(z - i) = 0$$

and whose points have a positive imaginary part.

Ans. (a) - (b) $y = -1/x$ (c) -

5 Show that one of the roots of the equation $(z - i)^5 = i(z - 1)^5$ is:

$$z = \frac{1}{2}\left(1 - \cot\frac{1}{20}\pi\right)(1 + i),$$

and find the other roots.

Hence show that $\cot\frac{1}{20}\pi + \cot\frac{5}{20}\pi + \cot\frac{9}{20}\pi + \cot\frac{13}{20}\pi + \cot\frac{17}{20}\pi = 5$

O

Ans. $z_2 = \frac{1}{2}(1 + i)\left(1 - \cot {}^{5\pi}/_{20}\right)$ where $k = 1$

$z_3 = \frac{1}{2}(1 + i)\left(1 - \cot {}^{9\pi}/_{20}\right)$ where $k = 2$

$z_4 = \frac{1}{2}(1 + i)\left(1 - \cot {}^{13\pi}/_{20}\right)$ where $k = 3$

$z_5 = \frac{1}{2}(1 + i)\left(1 - \cot {}^{17\pi}/_{20}\right)$ where $k = 4$.

6(a) In an Argand diagram the complex numbers p, q, r, are represented by the collinear points P, Q, R.

Prove that $\dfrac{q - p}{r - p}$ is real.

(b) In an Argand diagram, O is the origin, and the complex numbers a, b are represented by the points A, B.

Show that the area, Δ, of the triangle OAB (the area being regarded as positive if the order of vertices, as stated, is anti-clockwise) is given by

$$\Delta = \frac{1}{4}i\,(ab^* - a^*b).$$

Deduce that the perpendicular distance of the line AB from O is $\lambda|AB|$, where $\lambda = \dfrac{1}{2}i\dfrac{(ab^* - a^*b)}{|(a - b)|^2}$.

C

7(a) Show that, in the Argand diagram, the equation $|z + 3 + i| = 2|z + i|$ represents a circle.
Find the centre and radius of this circle.

(b) Solve the equation $z^3 + i = 0$, giving your answers in the form $re^{i\theta}$.

(c) Find the set of integers n for which $(1 + i)^n + (1 - i)^n = 0$.

C

Ans. (a) $C(1, -1)$, $r = 2$ (b) $e^{i\pi/2}$, $e^{i7\pi/6}$, $e^{i11\pi/6}$

(c) $n = 2(1 + 2k)$ where $k = 0, 1, 2 \ldots + (n - 1)$.

8 Given that $z = 2 - 3i$, express in the form $a + bi$ the complex numbers

(i) $(z + i)(z + 2)$ (ii) $\dfrac{z}{1 - z^2}$

Ans. (i) $2 - 14i$ (ii) $-\dfrac{2}{15} - \dfrac{7}{30}i$.

9 (i) Prove that $-1 + 2i$ is a root of $3x^3 + 5x^2 + 13x - 5 = 0$, and hence solve the equation completely.

(ii) Express $1 + \sqrt{3}i$ in the form $r(\cos\theta + i\sin\theta)$ where $r > 0$. Hence or otherwise, express $(1 + \sqrt{3}i)^9$ in the form $a + ib$ where a and b are real.

(iii) Find the real part of the complex number $\dfrac{1 + \cos\phi - i\sin\phi}{1 - \cos\phi - i\sin\phi}$, simplifying your answer as far as possible.

O.

Ans. (1) $-1 - 2i$, $x = {}^1{}_3$ (ii) $a = -2^9$, $b = 0$
(iii) $2\cos^2 \phi/2$.

10 (a) Find the modulus and argument of the complex number $\dfrac{3 - 2i}{1 - i}$.

(b) The fixed complex number a is such that $0 < \arg a < \frac{1}{2}\pi$.

In an Argand diagram, a is represented by the point A, and the complex number ia is represented by B. The variable complex number z is represented by P.

Draw a diagram showing A, B and the locus of P in each of the following cases:

(i) $|z - a| = |z - ia|$, (ii) $\arg(z - a) = \arg(ia)$.

Find, in terms of a, the complex number representing the point at which the loci intersect.

C.84

Ans. (a) $\sqrt{13/2}$, $11^\circ 18' 35''$
(b) (i) $y = \dfrac{x_1 + y_1}{y_1 - y_1}$ (ii) $yy_1 + xx_1 - y_1^2 - x_1^2 = 0$
$a + ia$.

11 (i) By using the result that $(\cos\theta + i\sin\theta)^n = \cos(n\theta) + i\sin(n\theta)$, or otherwise, show that

$$\tan 4\theta = \frac{4\tan\theta - 4\tan^3\theta}{1 - 6\tan^2\theta + \tan^4\theta}.$$

(ii) Obtain the roots of the equation $t^4 + 4t^3 - 6t^2 - 4t + 1 = 0$ giving your answers correct to two decimal places.

(iii) Given that α is not an integral multiple of $\frac{1}{4}\pi$, show that $\tan\alpha + \tan\left(\alpha + \frac{1}{4}\pi\right) + \tan\left(\alpha + \frac{1}{2}\pi\right) + \tan\left(\alpha + \frac{3}{4}\pi\right) = -4\cot 4a$.

C.84

Ans. (ii) 11.25°, 56.25°, 101.25°, 146.25°,
$t = 0.20$, 1.50, -5.03, -0.67

12 In the Argand diagram the two distinct circles with equations $zz^* - a_1^*z - a_1z^* + b_1 = 0$
and $zz^* - a_2^*z - a_2z^* + b_2 = 0$
intersect at the points A and B.

(i) Find the equation of the line AB in the form $az^* + a^*z + b = 0$.

(ii) Show that a necessary and sufficient condition for the tangents to the two circles at A to be perpendicular is $a_1a_2^* + a_1^*a_2 = b_1 + b_2$.

C.84

Ans. $az^* + a^*z + b = 0$.

13 Given that $z_1 = 3 + 2i$ and $z_2 = 4 - 3i$

(i) find $z_1 z_2$ and $\frac{z_1}{z_2}$, each in the form $a + ib$;

(ii) verify that $|z_1 z_2| = |z_1|\ |z_2|$. C.83

Ans. (i) $18 - i$, $\frac{6}{25} + \frac{17}{25}i$ (ii) -

14 Using De Moivre's Theorem for $(\cos\theta + i\sin\theta)^5$ or otherwise, prove that

$$\tan 5\theta = \frac{5\tan\theta - 10\tan^3\theta + \tan^5\theta}{1 - 10\tan^2\theta + 5\tan^4\theta}.$$

Prove that $\tan\frac{1}{20}\pi$ is a root of the equation $t^4 - 4t^3 - 14t^2 - 4t + 1 = 0$ and find the other roots in the form $\tan\frac{1}{20}n\pi$. C.84

Ans. $\tan n\pi/20$ where $n = 5, 9, 13$.

15 (a) Write down the modulus and argument of the complex number $1 + i$.
Hence, or otherwise, express $(1 + i)^5$ in the form $x + iy$, where x and y are real numbers.

(b) Sketch, on an Argand diagram, the locus given by
$z = 2 + \lambda(2 + i)$, where λ is a real parameter.

Sketch on the same diagram, the locus given by
$z = -3 + \mu(-1 + 2i)$, where μ is a real parameter.

Find by calculation the value(s) of z at the point(s) where these loci intersect. C.83

Ans. (a) $\sqrt{2}$, $\pi/4$, $-4 - 4i$. (b) $z = -2 - 2i$.

16 Write down the sum of the geometric series $z + z^2 + \ldots\ldots + z^n$.
Deduce by putting $z = e^{i\theta}$ in your result, or prove otherwise, that

$$\sin\theta + \sin 2\theta + \ldots. + \sin n\theta = \frac{\sin\frac{n\theta}{2}\sin\frac{(n+1)\theta}{2}}{\sin\frac{\theta}{2}}.$$

Hence, or otherwise, prove that

$$\sin\frac{\pi}{n} + \sin\frac{2\pi}{n} + \ldots\ldots + \sin\frac{(n-1)\pi}{n} = \cot\frac{\pi}{2n}.$$ C.83

17 Show, geometrically or otherwise, that for all complex numbers z and w

$$|z + w| \leqslant |z| + |w|.$$

State the relationship between arg z and arg w if the equality sign holds.

(i) The point P represents the complex number z in the Argand diagram.
Given that z varies so that $|z - 3| + |z + 3| = 10$ show that the $x - y$ equation of the locus of P is $16x^2 + 25y^2 = 400$.

(ii) Given that z is a complex number such that $|z - 3| + |z + 3| = 10$ prove that $|z| \leqslant 5$, and $|z - 3| \geqslant 2$. C.83

Ans. arg z = arg w

18 Define the complex conjugate z^* of the number $z = x + iy$ and show that zz^* is real.

Solve the equation $z^2 = z^*$.

Ans. $x = \frac{1}{2}$ $c(\frac{1}{2}, 0)$ $r = \frac{1}{2}$.

19 The complex number z satisfies each of the inequalities:

(1) $|z - 29 - 3i| \leqslant |z - 21 + 3i|$

(ii) $-\frac{1}{4}\pi \leqslant \arg z \leqslant \frac{1}{4}\pi$, (iii) $|z - 25| \leqslant 15$,

On a clearly labelled Argand diagram, indicate clearly the regions determined by each of these inequalities, and deduce that one of the inequalities is implied by another, identifying these two inequalities.

Find the set of possible values of $|z|$.
Find also the set of possible values of $\tan(\arg z)$.

Ans. (i) $4x + 3y \geqslant 100$, (ii) -

20 Express in the form $a + ib$

(i) $\dfrac{3 + 4i}{5 - 2i}$, (ii) $(\cos \pi/6 + i \sin \pi/6)^5$, (iii) $e^{i\pi/3}$.

Solve the equation $z^3 + 27 = 0$ and represent the roots on an Argand diagram.
The equation $z^3 + pz^2 + 40z + q = 0$ has a root $3 + i$, where p and q are real. Find the values of p and q.

Ans. (i) $\dfrac{7}{29} + \dfrac{18}{29}i$ (ii) $-\dfrac{\sqrt{3}}{2} + i\dfrac{1}{2}$ (iii) $\dfrac{1}{2} + i\dfrac{\sqrt{3}}{2}$

$z_1 = 3(\cos \pi/3 + i \sin \pi/3)$, $p = -11$
$z_2 = 3(\cos \pi + i \sin \pi)$, $q = -50$
$z_3 = 3(\cos 5\pi/3 + i \sin \pi/3)$

21 The roots of the equation $(z - 4)^3 = 8$ are z_1, z_2 and z_3. Find the roots and represent them by the points P_1, P_2, P_3 on an Argand diagram.
State the values of $|z_1|$, $|z_2|$ and $|z_3|$.

A complex number z is such that $|z| = k|z - 4|$.
Obtain the cartesian equation of the locus of z when
(i) $k = 1$, (ii) $k = 3$.
In each case sketch the locus.

For each of the points P_1, P_2 and P_3 find the value of k for which the point lies on the locus given by $|z| = k|z - 4|$.

Ans. $z_1 = 6$ $x = 2$ $c(^9/_2, 0)$ $r = {}^3/_2$
$z_2 = 3 + i\sqrt{3}$ $k = 3$, $k = \sqrt{3}$.
$z_3 = 3 - i\sqrt{3}$

22 The complex numbers z_1, z_2, z_3 are represented on an Argand diagram by the points A, B, C and

$$z_1 = 4i, \qquad z_2 = 2\sqrt{3} - 2i, \qquad z_3 = -2\sqrt{3} - 2i$$

(i) Show that the triangle ABC is equilateral.
(ii) Show that z_1^2 and $z_2 z_3$ are represented by the same point in the Argand diagram.
(iii) Find the equation of the circle OBC, where O is the origin in the $|z - a| = b$ form and also in the $f(x,y) = 0$ form.
(iv) Describe the locus of the point P, given that P represents the complex number z in the Argand diagram and

$$\arg\left(\frac{z - z_1}{z - z_2}\right) = \frac{\pi}{2}.$$

A.E.B.81

Ans. (i) - (ii) - (iii) $|z + 4i| = 4$

(iv) $x^2 + y^2 - 2x\sqrt{3} - 2y - 8 = 0$ $c(\sqrt{3},1)$ $r = 2\sqrt{3}$

(v) -

23 Solve the equation $z^3 = 8$ and show the three roots on an Argand diagram.
Find the non-real roots z_1 and z_2 of the equation

$$(z - 6)^3 = 8(z + 1)^3$$

expressing them in both cartesian and polar form.
Hence (a) show that $|z_1 - z_2| = 2\sqrt{3}$

(b) evaluate $z_1^5 + z_2^5$.

Prove that $\sum_{k=1}^{6} \exp(t_k \sqrt{2}) = 2 \cos 4 + 4 \cosh \sqrt{3} \cos 1$

where t_1 t_2 ....t_6 are the roots of the equation $(t^2 - 6)^3 = 8(t^2 + 1)^3$ and $\exp(x) \equiv e^x$.

A.E.B.83

Ans. $z_1 = 2$ $\qquad$ $z_1 = 1 + i\sqrt{3}$
$z_2 = 2(\cos 2\pi/3 + i \sin 2\pi/3)$ $\qquad$ $z_2 = 1 - i\sqrt{3}$
$z_3 = 2(\cos 4\pi/3 + i \sin 4\pi/3)$

(a) $2\sqrt{3}$ $\qquad$ (b) 32.

24(a) Show that the roots of the equation $z^3 = 1$ are 1, ω and ω^2, where

$$\omega = -\frac{1}{2} + \frac{i\sqrt{3}}{2}.$$

Express the complex number $5 + 7i$ in the form $A\omega + B\omega^2$, where A and B are real, and give the values of A and B in surd form.

(b) Given that $z = \cos\theta + i \sin\theta$, prove that $z^n + z^{-n} = 2 \cos n\theta$.
Hence prove that $\cos 5\theta = 16 \cos^5\theta - 20 \cos^3\theta + 5 \cos\theta$.

A.E.B.82

25 Write down in polar form, the five roots of the equation $z^5 = 1$.
Show that, when these five roots are plotted on an Argand diagram, they form the vertices of a regular pentagon of area $\frac{5}{2}\sin\frac{2\pi}{5}$.

By combining appropriate pairs of these roots, prove that for $z \neq 1$,

$$\frac{z^5 - 1}{z - 1} = \left(z^2 - 2z\cos\frac{2\pi}{5} + 1\right)\left(z^2 - 2z\cos\frac{4\pi}{5} + 1\right).$$

Use this result to deduce that $\cos\frac{2\pi}{5}$ and $\cos\frac{4\pi}{5}$ are the roots of the equation $4x^2 + 2x - 1 = 0$.

A.E.B.83

26 (a) Verify that $\alpha_1 = -1 - 3i$ is a root of the equation
$$z^2 + iz + 5(1 - i) = 0.$$

By considering the coefficient of z in the equation, or otherwise, find the second root α_2.

Find the modulus and argument of β where $\frac{1}{\beta} = \frac{1}{\alpha_1} + \frac{1}{\alpha_2}$.

(b) (i) Show that the locus of points in the Argand plane satisfying the equation $z\bar{z} + (1 + i)z + (1 - i)\bar{z} = 1$ is a circle.

(ii) Find the complex numbers corresponding to the points where the locus $z^2 + \bar{z}^2 - 14z\bar{z} + 48 = 0$ crosses the imaginary axis.

(iii) Show that the locus $2z^2 + 2\bar{z}^2 - z\bar{z} + 15 = 0$ does not cross the real axis.

W.J.E.C.82

Ans. (a) $\alpha_2 = 1 + 2i$ $|\beta| = \sqrt{50}$ arg $\beta = \pi/4$

(b) (i) $C(-1,1)$ $r = \sqrt{3}$
(ii) $12x^2 + 16y^2 = 48$ an ellipse.

27 Given that $\omega = \cos\frac{2\pi}{7} + i\sin\frac{2\pi}{7}$,
write down the modulus and argument of ω^4 and ω^5.

Plot the points represented by ω, ω^4 and ω^5 on an Argand diagram, and prove that they form the vertices of an isosceles triangle.

Write down the value of ω^7, and hence find the sum of the geometric series
$$1 + \omega + \omega^2 + \omega^3 + \omega^4 + \omega^5 + \omega^6.$$

Find the value of
$$(\omega + \omega^6)(\omega^2 + \omega^5) + (\omega^2 + \omega^5)(\omega^3 + \omega^4) + (\omega^3 + \omega^4)(\omega + \omega^6),$$

and hence find the cubic equation, with integer coefficients, whose roots are
$$(\omega + \omega^6), (\omega^2 + \omega^5) \text{ and } (\omega^3 + \omega^4).$$

A.E.B.84

28 Solve the equation: $z^2 - (4 - i)z + 9 + 7i = 0$, giving the roots in the form $a + ib$, with a and b real. Notice that the roots are not complex conjugates of each other.

Let $p(x)$ be a polynomial in x with real coefficients, and let α be a complex root of the equation $p(x) = 0$. Show that $\bar{\alpha}$, the conjugate of α, is also a root of this equation.

How do you reconcile this result with your answer to the first part of the question?

0.82

Ans. $z_1 = 3 - i2$ $z_2 = 1 + i3$

29 Find the modulus and argument of the complex number $\frac{1 - 3i}{1 + 3i}$.

Show that, as the real number t varies, the point representing $\frac{1 - it}{1 + it}$ in the Argand diagram moves round a circle, and write down the radius and centre of the circle. 0.82

Ans. 1, $-\tan^{-1} {}^{3}/_{1} - \tan^{-1} {}^{3}/_{1} = -2 \tan^{-1} 3$
$C(0,0)$, $r = 1$

30 By writing $2 \cos \theta = z + {}^{1}/z$, where $z = \cos \theta + i \sin \theta$, and applying De Moivre's theorem, show that

$$\cos^{2n-1} \theta = \left(\frac{1}{2}\right)^{2n-2} \left\{\cos(2n-1)\theta + \binom{2n-1}{1} \cos(2n-3)\theta + \ldots + \binom{2n-1}{n-1} \cos \theta\right\}$$

Hence or otherwise, evaluate $\int_0^{\pi/2} \cos^7\theta \, d\theta$ 0.83

Ans. -5/56

31 Show that, if z satisfies the equation

(a) $z^6 - 7z^4 + 7z^2 - 1 = 0$

then it also satisfies

(b) $(z + i)^8 = (z - i)^8$.

By solving (b), find the roots of the equation (a), and use these to find the values of

$\cot^2 \frac{1}{8}\pi + \cot^2 \frac{1}{4}\pi + \cot^2 \frac{3}{8}\pi$ and $\cot^2 \frac{1}{8}\pi \cot^2 \frac{1}{4}\pi \cot^2 \frac{3}{8}\pi$ 0.82

Ans. $z = \cot k\pi/8$, $\cot^2 \pi/8 + \cot^2 \pi/4 + \cot^2 3\pi/8 = 7$
$\cot^2 \pi/8 \ \cot^2 \pi/4 \ \cot^2 3\pi/8 = 1$

32 Let $z = 2(\sin \phi - i \cos \phi)$.
Express all the values of $z^{3/4}$ in the form $pe^{i\theta}$.
Show that they form the vertices of a square in the Argand diagram.
What is the length of the side of this square? 0.83

Ans. $2^{3/4} e^{-i(\pi/2 - \phi)3/4}$, $2^{3/4} e^{-i\{(\pi/2 - \phi)3 + 2\pi\}1/4}$,
$2^{3/4} e^{-i\{(\pi/2 - \phi)3 + 4\pi\}1/4}$, $2^{3\ 4} e^{-i\{(\pi\ 2 - \phi)3 + 6\pi\}^{1/4}}$
$2^{5/4}$ the side of the square.

33 Let z and w be complex numbers.
By using the modulus and argument forms of z and w, or otherwise, show that $\overline{z/w} = \bar{z}/\bar{w}$.
Deduce that if $z = \tan(u + iv)$, where u and v are real, then $\bar{z} = \tan(u - iv)$.

Show further that $Im\, z = \dfrac{\sinh 2v}{\cos 2u + \cosh 2v}$.

Finally show that, if $u = \frac{1}{8}\pi$ and v is allowed to vary, the locus of z in the Argand diagram is a circle whose centre is the point -1. Find the radius of the circle. O.83

34 Express $(6 + 5i)(7 + 2i)$ in the form $a + ib$.
Write down $(6 - 5i)(7 - 2i)$ in a similar form.
Hence find the prime factors of $32^2 + 47^2$. J.M.B.79

Ans. $32 + 47i$, $32 - 47i$

35 Indicate on an Argand diagram the region in which z lies, given that both $|z - (3 + i)| \leqslant 3$ and $\frac{\pi}{4} \leqslant \arg\{z - (1 + i)\} \leqslant \frac{\pi}{2}$ are satisfied

J.M.B.79

36 By using de Moivre's theorem, or otherwise, show that

$$\tan 5\theta = \frac{5t - 10t^3 + t^5}{1 - 10t^2 + 5t^4} \quad \text{where } t = \tan\theta,$$

Hence show that $\tan\frac{3\pi}{5} = -\sqrt{(5 + 2\sqrt{5})}$. J.M.B.75

37 Prove that the equation $z^3 - 3iz^2 - (9 + 4i)z - 8 - i = 0$ has a solution $z = 3 + 2i$.
Hence solve the equation completely, given that one of the other roots is real. J.M.B.76

Ans. $z = 3 + 2i$, $-2 + i$, -1.

38 Given that $z = i + e^{i\theta}$, show that $\left|\dfrac{2z - i}{iz - 1}\right|$ is independent of θ and state its value. Hence, or otherwise, show that the circle $|w - i/3| = 2/3$ in the w-plane is the image under the transformation $w = \dfrac{z - i}{iz - 1}$ of the circle $|z - i| = 1$ in the z-plane.

Ans. 1 J.M.B.78

39 Find the modulus and the argument of each of the roots of the equation $z^5 + 32 = 0$.
Hence express $z^4 - 2z^3 + 4z^2 - 8z + 16$ as the product of two quadratic factors of the form $z^2 - az\cos\theta + b$, where a, b and θ are real. J.M.B.77

Ans. $2/\pi/5$, $2/3\pi/5$, $2/5\pi/5$, $2/7\pi/5$, $2/9\pi/5$.

40 The roots of the quadratic equation $z^2 + pz + q = 0$ are $1 + i$ and $4 + 3i$.
Find the complex numbers p and q.

It is given that $1 + i$ is also a root of the equation $z^2 + (a + 2i)z + 5 + ib = 0$, where a and b are real.
Determine the values of a and b. J.M.B.78

Ans. $p = -5 - 4i$, $q = 1 + 7i$, $a = -3$, $b = -1$

41 Sketch the circle C with Cartesian equation $x^2 + (y-1)^2 = 1$.
The point P representing the non-zero complex number z, lies on C.
Express $|z|$ in terms of θ, the argument of z.
Given that $z^1 = \frac{1}{z}$, find the modulus and argument of z^1 in terms of θ.

Show that, whatever the position of P on the circle C, the point P^1 representing z^1 lies on a certain line, the equation of which is to be determined. J.M.B.78

Ans. $v = -\frac{1}{2}$

42 Given that $z = 4(\cos \pi/3 + i \sin \pi/3)$ and
$w = 2(\cos \pi/6 - i \sin \pi/6)$,
write down the modulus and argument of each of the following:

(i) z^3, (ii) $\frac{1}{w}$, (iii) $\frac{z^3}{w}$. J.M.B.78

Ans. (i) π (ii) $\pi/6$ (iii) $7\pi/6$

64 $\frac{1}{2}$ 16

43 Show in separate diagrams the regions of the z-plane in which each of the following inequalities is satisfied:

(i) $|z - 2| \leqslant |z - 2i|$,

(ii) $0 < \arg(z - 2) \leqslant \frac{\pi}{3}$.

Indicate clearly in each case, which part of the boundary of the region is to be included in the region.
Give the Cartesian equations of the boundaries. J.M.B.76

Ans. $y = \sqrt{3}x$, $y = 0$.

44 Shade in an Argand diagram the region of z-plane in which one or theother, but not both, of the following inequalities is satisfied:

(i) $|z| < 1$, (ii) $|z - 1 - i| \leqslant 2$.

Your diagram should show clearly which parts of the boundary are included. J.M.B.75

45 A transformation of the complex z-plane into the complex w-plane is given by

$$w = \frac{z - i}{2z + 1 + i}, \quad z \neq \frac{-(1+i)}{2}$$

(i) Prove that $z = \frac{w(1+i) + i}{1 - 2w}$, $w \neq \frac{1}{2}$

(ii) If $z = z^*$ prove that

$$ww^* + \frac{w}{4}(1 + i) + \frac{w^*}{4}(1 - i) - \frac{1}{2} = 0.$$

(iii) Hence, or otherwise, show that the real axis in the z-plane is mapped to a circle in the w-plane.
Give the centre and radius of this circle. J.M.B.79

Ans. $c\left(-\frac{1}{4}, \frac{1}{4}\right)$ $r = \frac{1}{2}\sqrt{\frac{5}{2}}$

46 (a) Solve the equation $z^6 - z^3 + 1 = 0$, giving your answers in the form $re^{i\theta}$.

(b) Find real numbers a, b, c, d such that

$$128 \cos^3\theta \sin^5\theta = a \sin 8\theta + b \sin 6\theta + c \sin 4\theta + d \sin 2\theta$$

for all values of θ.

Ans. $e^{i\pi/9}$ $e^{i7\pi/9}$ $e^{i13\pi/9}$, $e^{-i\pi/9}$, $e^{-i7\pi/9}$, $e^{-i13\pi/9}$,

$a = 1$, $b = -2$, $c = -2$, $d = 6$.

47 If $z = \cos\theta + i \sin\theta$, find $|z - 1|$ in its simplest form and show that $\arg(z - 1) = \frac{1}{2}(\pi + \theta)$. Hence find the arguments of the cube roots of $i - 1$ in terms of π. Find also the modulus of these cube roots to 3 significant figures.

If $z = x + iy$ is represented in an Argand diagram by the point P, sketch the locus of P when $|z| = 2|z - i + 1|$.

A.E.B.80

Ans. $|z - 1| = 2 \sin \theta/2$ $\arg(z - 1) = \frac{1}{2}(\pi + \theta)$, $\pi/4$, $11\pi/12$, $19\pi/12$

1.41 $c = (-4/3,\ 4/3)$, $r = 2\sqrt{2/3}$

48 Use De Moivre's theorem to show that

$$(1 + i \tan\theta)^n + (1 - i \tan\theta)^n = \frac{2 \cos n\theta}{\cos^n\theta}$$

Show that $i \tan \pi/8$ is a root of the equation

$$(1 + z)^4 + (1 - z)^4 = 0$$

and find the other three roots in symmetrical form.

Show that $\tan^2 \pi/8 = 3 - 2\sqrt{2}$.

MULTIPLE CHOICE QUESTIONS

1 The square root of 3 is a
a) Complex number
b) Real number
c) Negative complex number
d) Negative real number.

2 The square root of i^2 is a
a) Complex number
b) Real number
c) Rational number
d) Irrational number.

3 The roots of the quadratic equation $3x^2 - 5x + 3 = 0$ are
a) on the x-axis
b) on the y-axis
c) unequal real values
d) conjugate pairs of complex numbers

4 The straight line $2x + y - 3 = 0$ and the curve $x^2 = 3y$
a) intersect
b) touch
c) do not intersect
d) neet at infinity.

5 The imaginary part of the complex number $z = 3 - i4$ is
a) a positive real number
b) a negative real number
c) a positive complex number
d) a negative complex number.

6 The simplified expression for i^{1987} is
a) -1
b) 1
c) $-i$
d i

7 The sum of the complex numbers $z_1 = 3 - i4$, $z_2 = -3 + i4$, $z_3 = 1 - i5$ is
a) $1 + i5$
b) $1 - i5$
c) $0 - i13$
d) $7 + i13$

8 The conjugate of the complex number $z = -2 - i7$ is
a) $2 + i7$
b) $-2 - i7$
c) $-2 + i7$
d) $2 - i7$

9 The modulus of the complex number $z = \dfrac{\sqrt{2}}{1 - i}$ is
(a) $2^{1/4}$ (b) $2^{\frac{1}{2}}$ (c) 1 (d) $1/\sqrt{2}$

10 The argument of the complex number $z = \dfrac{\sqrt{3}}{1 + i\sqrt{3}}$ is

(a) $\tan^{-1} \dfrac{\sqrt{3}}{1}$ (b) $-\tan^{-1} \dfrac{\sqrt{3}}{1}$

(c) 0 (d) $180^\circ - \tan^{-1} \sqrt{3}$

11 The polar form of the complex number = 5 - i3 is

(a) $\sqrt{34}\ \underline{/180^\circ + \tan^{-1} 3/5}$ (b) $\sqrt{34}\ \underline{/180^\circ - \tan^{-1} 3/5}$

(c) $\sqrt{34}\ \underline{/360^\circ - \tan^{-1} 3/5}$ (d) $\sqrt{34}\ \underline{/\tan^{-1} 3/5}$

12 The simplified expression for $\dfrac{3\underline{/-45^\circ}}{5\underline{/-90^\circ}}$ is

(a) $0.6\underline{/45^\circ}$ (b) $0.6\underline{/-45^\circ}$ (c) $-0.6\underline{/45^\circ}$ (d) $0.6\underline{/135^\circ}$

13 The exponential form of the complex number $5\underline{/-35^\circ}$ is

(a) $5e^{i35}$ (b) $5e^{-i35\pi}$

(c) $5e^{-35i}$ (d) $5e^{-i0.611}$

14 The complex number $7e^{-i\pi/2}$ is written as

(a) $7\underline{/\pi/2}$ (b) $7\underline{/-\pi/2}$ (c) $7\cos\pi/2 + i7\sin\pi/2$
(d) $-7\cos\pi/2 - i7\sin\pi/2$

15 The square root of $\sqrt{1 + i}$ is

(a) $1.01 - i0.46 = 1.11\ \underline{/-24^\circ 29'}$

(b) $1.01 + i0.46 = 1.11\ \underline{/\ 24^\circ 29'}$

(c) $-1.01 - i0.46 = 1.11\ \underline{/180^\circ - 24^\circ 29'}$

(d) $-1.01 + i0.46 = 1.11\ \underline{/180^\circ - 24^\circ 29'}$

16 The simplified expression for $= \underline{/\pi/4} \cdot \underline{/\pi/3} \cdot \underline{/\pi/2} \cdot \underline{/\pi/1}$ is

(a) $\underline{/\pi/12}$ (b) $\underline{/-\pi/12}$ (c) $e^{-i\pi/12}$ (d) $-e^{i\pi/12}$

17 The coefficient of $\sin^7\theta\ \cos^3\theta$ in the expansion $(\cos\theta + i\sin\theta)^{10}$ is

(a) $120i$ (b) $-120i$ (c) 120 (d) -120

18 If $z = x + iy$ then $\overline{\left(\frac{1}{z}\right)}$ is equal to

(a) $\dfrac{1}{x + iy}$ (b) $\dfrac{1}{x - iy}$

(c) $\dfrac{x + iy}{1}$ (d) $\dfrac{x}{x^2 + y^2} - i\,\dfrac{y}{x^2 + y^2}$

19 If $z = x + iy$ and $z^* = x - iy$ then zz^* is equal to

(a) $\dfrac{1}{x^2 + y^2} = \dfrac{1}{|z|^2}$ (b) $\dfrac{1}{x^2 - y^2}$

(c) $x^2 + y^2 = |z|^2$ (d) $x^2 - y^2$

20 If z_1, z_2 are any complex numbers, then

(a) $\overline{z_1 + z_2} = \overline{z_1} - \overline{z_2}$ (b) $\overline{z_1 + z_2} = z_1 + z_2$

(c) $z_1 + z_2 = \overline{z_1} + \overline{z_2}$ (d) $\overline{z_1 + z_2} = \overline{z_1} + \overline{z_2}$

21 The value of $\dfrac{e^{i\theta} + e^{-i\theta}}{2}$ is
(a) $\sinh\theta$
(b) $\cosh i\theta$
(c) $\cosh\theta$
(d) $\sinh i\theta$

22 The value of $e^{i\pi/2} - e^{-i\pi/2}$ is
(a) $2\sinh\pi/2$
(b) $2\sin\pi/2$
(c) $2\cosh\pi/2$
(d) $2\cos\pi/2$

23 The value of $\cosh\pi$ is
(a) $\cos\pi$
(b) $\cos i\pi$
(c) $\sin\pi$
(d) $i\sinh\pi$

24 The $\log_e(-2)$ is
(a) $\ln 2 + i\pi$
(b) $\ln 2 - i\pi$
(c) $-\ln 2 - i\pi$
(d) $-\ln 2 + i\pi$

25 The roots of $z^3 = 7 + 24i$ are

(a) $25^{1/3}\ \underline{/24^{\circ}35'}$, $25^{1/3}\ \underline{/144^{\circ}35'}$ and $25^{1/3}\ \underline{/264^{\circ}35'}$

(b) $25^{1/3}\ \underline{/155^{\circ}25'}$, $25^{1/3}\ \underline{/275^{\circ}25'}$ and $25^{1/3}\ \underline{/35^{\circ}25'}$

(c) $25^{1/3}\ \underline{/204^{\circ}35'}$, $25^{1/3}\ \underline{/324^{\circ}35'}$ and $25^{1/3}\ \underline{/84^{\circ}35'}$

(d) $25^{1/3}\ \underline{/-24^{\circ}35'}$, $25^{1/3}\ \underline{/-144^{\circ}35'}$ and $25^{1/3}\ \underline{/-264^{\circ}35'}$

26 The roots of the cubic equation $z^3 - 1 = 0$ are

(a) $1,\ \frac{1}{2} + i\frac{\sqrt{3}}{2},\ \frac{1}{2} - i\frac{\sqrt{3}}{2}$

(b) $-1,\ -\frac{1}{2} - i\frac{\sqrt{3}}{2},\ \frac{1}{2} + i\frac{\sqrt{3}}{2}$

(c) $1,\ -\frac{1}{2} + i\frac{\sqrt{3}}{2},\ -\frac{1}{2} - i\frac{\sqrt{3}}{2}$

(d) $-1,\ -\frac{1}{2} + i\frac{\sqrt{3}}{2},\ -\frac{1}{2} + i\frac{\sqrt{3}}{2}$

27 The locus of $|z + i| = 1$ is
(a) a straight line
(b) a circle with centre $C\ (1,1)$ and radius $r = 1$
(c) the cartesian equation $x^2 + y^2 + 2y = 0$
(d) a circle passing through the point $(0,1)$.

28 The locus of the point representing z, where $\left|\frac{z - 1}{z + 1}\right| = \frac{1}{2}$ is a circle with coordinates and radius respectively

(a) $(-5/3,\ 0)$ $r = 4/3$

(b) $(0,\ 5/3)$, $r = 4/3$

(c) $(5/3,\ 0)$, $r = 4/3$

(d) $(5/3\ 5/3)$ $r = 4/3$

29 The square roots of $2i$ are

(a) $\sqrt{2}\underline{/\pi/4}$, $\sqrt{2}\underline{/-\pi/4}$

(b) $\sqrt{2}\underline{/\pi/4}$, $\sqrt{2}\underline{/3\pi/4}$

(c) $\sqrt{2}\underline{/\pi/4}$, $\sqrt{2}\underline{/5\pi/4}$

(d) $\sqrt{2}\underline{/\pi/4}$, $\sqrt{2}\underline{/\pi}$

30 The square roots of $-2i$ in the form $\pm(a + ib)$ are

(a) $-1 + i1,\quad 1 - i1$

(b) $1 + i1,\quad -1 - i1$

(c) $-1 - i1,\quad -1 + i1$

(d) $1 + i1,\quad 1 - i1$

31 The complex form of a circle with centre $C(-1, -2)$ and radius $r = 2$, is written as

(a) $|z + 1 + 2i| = 2$

(b) $|z - 1 - 2i| = 2$

(c) $|z - 1 + 2i| = 2$

(d) $|z + 1 - 2i| = 2$

32 The complex form of a circle is $|z - i| = 3$, then the circle has the following properties:

(a) $C(-1,1)$, $r = 3$

(b) $C(0,1)$, $r = 3$

(c) $C(-1, -1)$ $r = 3$

(d) $C(1, 1)$, $r = 3$

33 If $z = x + iy$ is represented by the point $P(x,y)$ in the z-plane and $w = u + vi$ is represented by the point $Q(u,v)$, in the w-plane, then the relationship betweeen z and w, $zw = 2$, defines the circle $|z| = 5$ of the point P, which is mapped onto the point Q as

(a) $|w| = 2/5$

(b) $u^2 + v^2 = 2/5$

(c) $C(0,0)$ $r = 4/25$

(d) $C(0,0)$, $r = 2$

34 The roots of the quadratic equation $z^2 - 4z + 8 = 0$ are

(a) unequal and real

(b) equal and real

(c) complex and equal

(d) complex and conjugate

35 The locus of $\arg \frac{z - 1}{z + 1} = \pi/4$ is a

(a) parabola which cuts the y-axis at ± 1

(b) parabola which cuts the y-axis at $+1$

(c) parabola which cuts the x-axis at ± 1

(d) parabola which has a maximum at $(0, -1)$

RECAPITULATION OR SUMMARY

$z = x + jy = r(\cos\theta + j\sin\theta) = re^{j\theta}$

r = modulus

θ = argument or amplitude

$x = R(z)$

$y = I(z)$

$r = |z|$

$\theta = \arg z$

$\bar{z} = x - jy = re^{-j\theta}$ conjugate

The point representing $\bar{z}$ is the reflection of the point in the real axis.

$x = \frac{1}{2}(z + \bar{z}), \quad |z| = r = \sqrt{z\bar{z}}$

$y = \frac{1}{2}j(z - \bar{z})$

If $z_1 = x_1 + jy_1 = r_1 e^{j\theta_1}, \quad z_2 = x_2 + jy_2 = r_2 e^{j\theta_2}$

$$z_1 \pm z_2 = (x_1 \pm x_2) + j(y_1 \pm y_2)$$

$$z_1 z_2 = r_1 r_2 e^{j(\theta_1 + \theta_2)}$$

$$|z_1 z_2| = |z_1||z_2|$$

$$\arg(z_1 z_2) = \arg z_1 + \arg z_2$$

$$\frac{z_1}{z_2} = \frac{r_1}{r_2} e^{j(\theta_1 - \theta_2)}$$

$$\frac{|z_1|}{|z_2|} = \frac{|z_1|}{|z_2|}, \quad \arg\frac{z_1}{z_2} = \arg z_1 - \arg z_2$$

$$z^n = r^n e^{jn\theta} = r^n e^{jn(\theta + 2k\pi)}$$

$$= r^n \cos(n\theta + k.2n\pi) + j\sin(n\theta + k.2n\pi)$$

where k is an integer.

If n is a fraction $n = p/q$, there are q distinct values of z^n, corresponding to $k = 0, 1, 2 \ldots\ldots$

Inequalities

$$|z_1 + z_2| \leq |z_1| + |z_2|$$

$$e^{iz} = 1 + iz + \frac{(iz)^2}{2!} + \frac{(iz)^2}{3!} + \ldots\ldots$$

$$\log_e z = \log_e(re^{j\theta}) = \log_e r + j\theta$$

$$e^z = 1 + z + \frac{z^2}{2!} + \frac{z^3}{3!} + \frac{z^4}{4!} + \ldots\ldots$$

$$\sin z = z - \frac{z^3}{3!} + \frac{z^5}{5!} - \ldots\ldots$$

$$\cos z = 1 - \frac{z^2}{2!} + \frac{z^4}{4!} - \ldots\ldots$$

Answers to MULTIPLE CHOICE questions:

1 (b)	2 (a)	3 (d)	4 (a)	5 (b)
6 (c)	7 (b)	8 (c)	9 (c)	10 (b)
11 (c)	12 (a)	13 (d)	14 (b)	15 /
16 (a)	17 (b)	18 (b)	19 (c)	20 (d)
21 (b)	22 (a)	23 (b)	24 (a)	25 (a)
26 (c)	27 (c)	28 (c)	29 (c)	30 (a)
31 (a)	32 (b)	33 (a)	34 (d).	35 /

INDEX

Books in Preparation

TITLES IN THE BTEC SERIES

- MATHEMATICS II
- ANALYTICAL MATHEMATICS II
- ALGEBRA III
- CALCULUS III
- MATHEMATICS III
- ELECTRICAL AND ELECTRONIC PRINCIPLES II
- ELECTRICAL AND ELECTRONIC PRINCIPLES III
- ELECTRONICS II
- ELECTRONICS III